Algebraic Numbers and Algebraic Functions

Algebraic Numbers and Algebraic Functions

P. M. Cohn, FRS

Department of Mathematics
University College London

CHAPMAN & HALL
London · New York · Tokyo · Melbourne · Madras

Published by Chapman & Hall, 2–6 Boundary Row, London SE1 8HN

Chapman & Hall, 2–6 Boundary Row, London SE1 8HN, UK

Chapman & Hall 29 West 35th Street, New York NY10001, USA

Chapman & Hall Japan, Thomson Publishing Japan, Hirakawacho Nemoto Building, 7F, 1-7-11 Hirakawa-cho, Chiyoda-ku, Tokyo 102, Japan

Chapman & Hall Australia, Thomas Nelson Australia, 102 Dodds Street, South Melbourne, Victoria 3205, Australia

Chapman & Hall India, R. Seshadri, 32 Second Main Road, CIT East, Madras 600 035, India

First edition 1991

Typeset in 10/12 Times by Pure Tech Corporation, India
Printed in Great Britain by T.J. Press (Padstow) Ltd, Padstow, Cornwall.

ISBN 0 412 36190 6

A catalogue record for this book is available from the British Library

Library of Congress Cataloging-in-Publication data available

Contents

Preface

One of the most interesting and central areas of mathematics is the theory of algebraic functions – algebra, analysis and geometry meet here and interact in a significant way. Some years ago I gave a course on this topic in the University of London, using Artin's approach via valuations, which allowed one to treat algebraic numbers and functions in parallel. At the invitation of my friend Paulo Ribenboim I prepared a set of lecture notes which was issued by Queen's University, Kingston, Ontario, but I had always felt that they might be deserving of a wider audience.

Ideally one would first develop the algebraic, analytic and geometric background and then pursue the theme along these three paths of its development. This would have resulted in a massive and not very readable tome. Instead I decided to assume the necessary background from algebra and complex analysis and leave out the geometric aspects, except for the occasional aside, but to give a fairly full exposition of the necessary valuation theory. This allowed the text to be kept to a reasonable size.

Chapter 1 is an account of valuation theory, including all that is needed later, and it could be read independently as a concise introduction to a powerful method of studying general fields. Chapter 2 describes the behaviour under extensions and shows how Dedekind domains can be characterized as rings of integers for a family of valuations with the strong approximation property; this makes the passage to extensions particularly transparent. These methods are then put to use in Chapter 3 to characterize global fields by means of the product formula, to classify the global fields and to prove two basic results of algebraic number theory in this context: the unit theorem and the finiteness of the class number.

Chapter 4, the longest in the book, treats algebraic function fields of one variable. The main aim is the description of the group of divisor classes of degree zero as a particular $2g$-dimensional compact abelian group, the Jacobian variety. This is the Abel–Jacobi theorem; the methods used are as far as possible algebraic, although the function–theoretic interpretation is borne in mind. On the way automorphisms of function fields are discussed and the special case of elliptic function fields is developed in a little more detail. A final brief chapter examines the case of valuations on fields of two variables; in a sense this is a continuation of Chapter 1 which it is hoped will illuminate the development of Chapter 4. The main exposition follows the notes, but the telegraphese style has been expanded and clarified where necessary, and there have been several substantial additions, especially in Chapter 4.

The prerequisites are quite small: an undergraduate course in algebra and in complex function theory should suffice; references are generally given for any results needed. There are a number of exercises containing examples and further developments.

In writing a book of this kind one inevitably has a heavy debt to others. The first three chapters were much influenced by the writings of E. Artin quoted in the bibliography. A course on number theory by J. Dieudonné (which I attended at an impressionable age) has helped me greatly in Chapter 2. The treatment of valuations on function fields of two variables (Chapter 5) is based on a paper by Zariski, while Chapter 4 owes much to several works in the bibliography, particularly those by Hensel and Landsberg, Tchebotarev and by Eichler. I should like to thank Paulo Ribenboim for bringing out the earlier set of notes and a number of friends and colleagues for their advice and help, particularly David Eisenbud, Frank E. A. Johnson and Mark L. Roberts. The latter has also helped with the proof reading. Further I am indebted to the staff of Chapman & Hall for the way they have accommodated my wishes.

P. M. Cohn
University College London
May 1991

Notes to the reader

This book is intended for readers who have a basic background in algebra (elementary notions of groups, rings, fields and linear algebra). Occasionally results from Galois theory or commutative ring theory are used, but in such cases references are usually given, mostly to the author's algebra text, referred to as A.1, 2, 3 (see the bibliography). For Chapter 4 an acquaintance with the elements of complex function theory will be helpful.

Using notation which, thanks to Bourbaki, has become standard, we write **N**, **Z**, **Q**, **R**, **C** for the natural numbers, integers, rational, real and complex numbers respectively and $\mathbf{F}_p$ for the field of p elements (integers mod p). Rings are always associative, with a unit element, written 1, and except for some occasions in Chapter 1 they are commutative. If $1 = 0$, the ring is reduced to 0 and is called *trivial*; mostly our rings are non-trivial. If K is any ring, the polynomial ring in an indeterminate x over K is written $K[x]$; when K is a field, $K[x]$ has a field of fractions, called the *field of rational functions* in x over K and written $K(x)$. A polynomial is called *monic* if its leading coefficient is 1. A ring R is called a *K-algebra* if it is a K-module and the multiplication in R is K-linear in each argument, e.g. $K[x]$ is a K-algebra. The additive group of a ring R is denoted by R^+ and the set of non-zero elements in R by $R^\times$; if this set contains 1 and is closed under multiplication, R is called an *integral domain*. In a group G the subgroup generated by a subset X is denoted by $\langle X \rangle$. If H is a subgroup of G, a *transversal* of H in G is a complete set of representatives (for the cosets of H in G). A set with an associative binary operation is a *semigroup*; if it also contains a neutral element for the multiplication (e such that $ex = xe = x$ for all x) it is called a *monoid*.

In one or two places Zorn's lemma is used; we recall that this is the algebraist's version of the axiom of choice. It states that a partially ordered set A has a maximal element provided that every totally ordered subset has an upper bound in A. If every totally ordered subset in A has an upper bound, A is called *inductive*; so the lemma states that every inductive partially ordered set has a maximal element (cf. A.2, section 1.2 for a discussion). A property of points of a set is said to hold at *almost all* points if it holds for all except a finite number of points.

In the text, the pth item (Theorem, Proposition, Corollary or Lemma) in section $m.n$ is referred to as item $n.p$ within Chapter m and as item $m.n.p$ elsewhere. Thus, the first item in section 1.3 (in this case, a Theorem) is referred to as item 3.1 within Chapter 1 and as item 1.3.1 elsewhere within the book.

Introduction

The starting point for an algebraic treatment of number theory is the ring **Z** of integers, basic for so much of mathematics. A fundamental property which facilitates its study is the Euclidean algorithm; as an easy consequence one obtains the unique factorization of integers (the 'fundamental theorem of arithmetic') and the fact that **Z** is a principal ideal domain. These properties no longer always hold for the ring of integers in an algebraic number field; as Dedekind found, instead of the unique factorization of integers one merely has the unique factorization of ideals – in current terminology one is dealing with a *Dedekind domain.* More generally, a finite integral extension of a principal ideal domain is a Dedekind domain (provided it is integrally closed), and starting from the polynomial ring over a field, $k[x]$, one so obtains rings of functions of one variable, or geometrically speaking, rings of functions on an algebraic curve.

In this way the theory of Dedekind rings can be used to study rings of algebraic integers and rings of algebraic functions of one variable: the many parallels which exist here make it natural to develop a theory that can be applied in both cases. This is done in Chapter 2, after a preliminary chapter on valuations, which form a convenient means of studying Dedekind domains. The case of number fields is treated in Chapter 3, in a slightly more general form (global fields), which allows the application to function fields to be made with minimal modifications. Chapter 4 is devoted to function fields; this is an extraordinarily rich area, a meeting point of algebra, function theory and geometry, but consonant with the character of this book we have limited ourselves to the more immediate consequences.

What is perhaps the main aim in Chapters 3 and 4 may be described as follows: let A be a Dedekind domain with field of fractions K. Each non-zero element α of K gives rise to a principal fractional ideal (α) and the mapping $\phi : \alpha \mapsto (\alpha)$ can be described by the exact sequence

$$1 \to U \to K^{\times} \xrightarrow{\phi} F(A) \to C(A) \to 1,$$

where $U = \ker \phi$ is the group of units of A, $F(A)$ is the group of fractional ideals and $C(A) = \operatorname{coker} \phi$ is the ideal class group.

For a general Dedekind domain not much more can be said; thus $C(A)$ can be an arbitrary abelian group (Claborn [1966]), but in the cases of interest to us more information is available.

1. When A is the ring of integers in an algebraic number field K, then the unit group U is a finitely generated abelian group with cyclic torsion group and of rank $r_1 + r_2 - 1$, where r_1 is the number of real and $2r_2$ the number of complex conjugates of K (Dirichlet's unit theorem; see section 3.3); the ideal class group $C(A)$ is a finite group, as is proved in section 3.4; more precise estimates for the class number, i.e. the order of $C(A)$, can be obtained by analytic methods, cf. Borevich–Shafarevich (1966).
2. In the case where K is an algebraic function field of one variable, U is just the subfield of constants. Now the fractional ideals are described by divisors and it is convenient to replace F by F_0, the subgroup of divisors of degree zero (cf. section 4.1); then $\operatorname{coker} \phi \cong K^{\times}/F_0(A)$ is a compact abelian group, the *Jacobian variety*. This is the content of the Abel–Jacobi theorem (cf. section 4.8), and it ends our main story.

The final chapter is a brief description of function fields of two variables. While a general theory of such fields is beyond the scope of this book (cf. Jung (1951); Schafarewitsch (1968)), this classification helps to illuminate the one-dimensional case, and it forms a good illustration of the way valuations can be used in this study.

1

Fields with valuations

The notion of valuation is basic in the study of fields and will be a useful tool in later chapters. We therefore begin by introducing it, concentrating mainly on the properties needed later, but placing them in a wider context. After defining absolute values, which generalize an idea familiar from real numbers, and exploring the topological and metric consequences, we come to the main topic of this chapter in section 1.4, where the triple correspondence: valuations, valuation rings, places is explained and illustrated. The structure of complete fields is further elucidated by the power series representation, to which section 1.5 is devoted.

In the last two sections 1.6 and 1.7 we briefly examine general valuations. These results are only used at one place (Chapter 5), and these sections may be omitted on a first reading, but the reader is urged to return to them later, since they help to place the more commonly used rank 1 valuations in perspective.

1.1 ABSOLUTE VALUES

Let R be a non-trivial ring; by an *absolute value* on R we understand a real-valued function $x \mapsto |x|$ on R such that

A.1 $|x| \geqslant 0$, with equality if and only if $x = 0$;

A.2 $|x + y| \leqslant |x| + |y|$ (triangle inequality);

A.3 $|xy| = |x| . |y|$.

We note some easy consequences of the definitions. If R is a ring with an absolute value $|\ |$, then since $R \neq 0$, the absolute value is not identically zero by **A**.1, so $|1| = 1$ by **A**.3 and hence $|-1|^2 = |1| = 1$. By **A**.1 it follows that $|-1| = 1$, and more generally,

$$|-x| = |x| \quad \text{for all } x \in R. \tag{1}$$

Now **A**.3 with **A**.1 shows that the product of non-zero elements must be non-zero, so that R is an integral domain; in fact we shall mainly be concerned with fields.

Further, using **A**.2, we find that $|x-y| \leqslant |x|+|y|$ and as in elementary analysis we deduce that $||x|-|y|| \leqslant |x+y|$. From **A**.3 we also find, by induction, that

$$|x_1+\ldots+x_n| \leqslant |x_1|+\ldots+|x_n|.$$

In a field, **A**.2 can be replaced by

$$|1+z| \leqslant 1+|z|. \tag{2}$$

For this is a special case of **A**.2, and we regain **A**.2 by putting $z=xy^{-1}$ and multiplying by $|y|$, using **A**.3 (this argument assumes that $y \neq 0$, but of course **A**.2 holds trivially when $y=0$).

Let us list some examples:

(i) An obvious example is the usual absolute value on the real numbers **R** or the complex numbers **C**. If we interpret complex numbers as vectors in the plane, then $x, y, x+y$ correspond to the vertices of a triangle, and **A**.2 just expresses the fact that no side of a triangle can be greater than the sum of the other two.

(ii) If we fix a prime number p, any non-zero rational number c can be written in the form

$$c=p^{\nu}\cdot\frac{m}{n}, \quad \text{where } \nu \in \mathbf{Z}, m \in \mathbf{Z}, n \in \mathbf{N}, p \nmid mn. \tag{3}$$

Here the integer ν, indicating how often p divides c, is uniquely determined by c. We now put

$$|c|=2^{-\nu}, \quad \text{where } \nu \text{ is as in (3).} \tag{4}$$

This defines an absolute value on **Q**, as is easily checked. It is called the *p-adic value* and was first introduced by Hensel in 1904 (probably guided by the analogy to the case of function fields, cf. Hensel and Landsberg, (1902)). The number 2 on the right of (4) can clearly be replaced by any number greater than 1.

(iii) Let k be any field and in the polynomial ring $k[x]$ choose a polynomial $p(x)$ which is irreducible over k. Then any non-zero rational function ϕ in x over k can be written as

$$\phi=p^{\nu}\cdot\frac{f}{g}, \quad \text{where } \nu \in \mathbf{Z}, f, g \in k[x], p \nmid fg.$$

We obtain an absolute value on the rational function field $k(x)$ by setting

$$|\phi| = 2^{-\nu}.$$

For example, if k is algebraically closed, then p is linear and so may be taken in the form $p = x - \alpha$ ($\alpha \in k$). Here ν indicates the *order* of the function ϕ at the point $x = \alpha$. It is a *zero* of ϕ if $\nu > 0$ and a *pole* if $\nu < 0$.

(iv) The function

$$|x| = \begin{cases} 0 & \text{if } x = 0, \\ 1 & \text{if } x \neq 0, \end{cases}$$

defines an absolute value on any integral domain. It is called the *trivial* absolute value; here the restriction to integral domains is necessary to ensure that **A**.3 is satisfied.

We observe that the cases listed in (ii)–(iv) satisfy the following stronger form of **A**.2:

A.2′ $\quad |x + y| \leqslant \max\{|x|, |y|\}$ (ultrametric inequality).

An absolute value satisfying **A**.2′ is called *non-archimedean*, any other is called *archimedean*. For a non-archimedean absolute value we have

$$|x - y| = \max\{|x|, |y|\} \quad \text{whenever } |x| \neq |y|. \tag{5}$$

For we have $|x - y| \leqslant \max\{|x|, |y|\}$ by **A**.2′; if this inequality is strict and $|x| \neq |y|$, say $|x| < |y|$, then $|y| > \max\{|x - y|, |x|\}$, which contradicts **A**.2′, because $y = x - (x - y)$. This proves equation (5); it may be more imaginatively expressed by saying that for a non-archimedean absolute value all triangles are isosceles.

From **A**.2′ and (5) we find by induction on n,

$$|x_1 + \ldots + x_n| \leqslant \max\{|x_1|, \ldots, |x_n|\},$$

with strict inequality holding only if at least two of the $|x_i|$ are equal.

For any absolute value, if $u^n = 1$ for some $n > 0$, then $|u|^n = 1$, hence $|u| = 1$. Since in a finite field of q elements, every non-zero element x satisfies the equation $x^{q-1} = 1$, it follows that a finite field has no absolute value apart from the trivial one. This holds more generally for algebraic extensions of finite fields; later we shall see that every field not algebraic over a finite field has a non-trivial absolute value.

Generally speaking, absolute values on (commutative) rings are in some sense equivalent to absolute values on fields. More precisely, every absolute value on a commutative ring can be extended to an absolute value on its field of fractions, and to define an absolute value on a field K it is enough to define it on a subring having K as field of fractions.

Proposition 1.1
Let R be a commutative ring with an absolute value $|\ |$. Then R is an integral domain and the absolute value can be extended in just one way to an absolute value on the field of fractions of R.

Proof
We have seen that R must be an integral domain. Let K be its field of fractions; any $u \in K$ has the form $u = ab^{-1}$, where $a, b \in R, b \neq 0$, so if $|\ |$ can be extended to K, then since $a = ub$, we have by **A**.3,

$$|a| = |ub| = |u| . |b|. \tag{6}$$

This shows that there is at most one extension. To show that one exists we simply check that the value for $|u|$ given by (6) is indeed an absolute value. If also $u = a_1 b_1^{-1}$, then $ab_1 = ba_1$, hence

$$|a| . |b_1| = |b| . |a_1|$$

and so

$$|a_1| . |b_1|^{-1} = |a| . |b|^{-1};$$

this shows that $|\ |$ is well-defined on K. The verification of **A**.1–**A**.3 is straightforward and may be left to the reader.■

We also note the various ways of characterizing the non-archimedean property:

Proposition 1.2
For any absolute value $|\ |$ on a commutative ring R the following conditions are equivalent:

(a) $|\ |$ *is non-archimedean;*
(b) $|m . 1| \leqslant 1$ *for all* $m \in \mathbf{N}$*;*
(c) $|m . 1| \leqslant 1$ *for some* $m > 1$*;*
(d) $|m . 1|$ *is bounded for* $m \in \mathbf{N}$.

Proof
(*a*) ⇒ (*b*) ⇒ (*c*) are clear, so assume (c). To prove that (d) holds we shall need the elementary limit:

$$\lim_{n \to \infty} (1 + n\alpha)^{1/n} = 1 \quad \text{for any real } \alpha \geqslant 0. \tag{7}$$

It follows by first observing that $(1 + n\alpha)^{1/n} \geqslant 1$; now for any $\delta > 0$ we have $(1 + \delta)^n > 1 + n\alpha$, for all sufficiently large n, hence

$$1 + \delta > (1 + n\alpha)^{1/n} \geqslant 1,$$

and now (7) follows since δ was arbitrary.

Given (c), we take any $n \in \mathbf{N}$ and express it in the base m:

$$n = a_0 + a_1 m + \ldots + a_r m^r, \quad \text{where } 0 \leqslant a_i < m, a_r \neq 0.$$

It follows that $m^r \leqslant n$, hence $r \leqslant \log n$, where the logarithm is taken to base m. Thus $|n| \leqslant m(1 + \log n)$. Replacing n by n^s we obtain

$$|n|^s \leqslant m(1 + s . \log n).$$

We now take sth roots and let $s \to \infty$. Using (7), we find

$$|n| \leqslant \lim_{s \to \infty} (m(1 + s . \log n))^{1/s} = 1,$$

and this proves (d).

Now assume (d): there is a constant M such that $|n . 1| \leqslant M$ for all n; hence

$$|x + y|^n = |(x + y)^n| = \left|\sum \binom{n}{i} x^i y^{n-i}\right| \leqslant M . \sum |x|^i |y|^{n-i}.$$

If $|x| \leqslant |y|$, say, then $|x|^i |y|^{n-i} \leqslant |y|^n$, so it follows that

$$|x + y|^n \leqslant (n + 1) M . \max\{|x|^n, |y|^n\}.$$

Taking nth roots and letting $n \to \infty$, we find (using (7)) that

$$|x + y| \leqslant \max\{|x|, |y|\},$$

i.e. (a) holds.■

Corollary 1.3
Any field with an archimedean absolute value has characteristic zero.

For in characteristic $p \neq 0$ the set $\{|n . 1| \mid n \in \mathbf{N}\}$ is finite and hence bounded.■

Exercises

1. Define a *norm* on a ring R as a real-valued function $x \mapsto |x|$ satisfying **A**.1 and **A**.2 and $|xy| \leqslant |x| . |y|$. Show that every non-trivial ring has a norm; show further that $|1| \geqslant 1$ and if c is invertible, then $|c^{-1}| \geqslant |c|^{-1}$.
2. Verify that the examples given in (i)–(iv) are absolute values. Show also that the function defined on $k(x)$ by

$$|f/g| = 2^{deg f - deg g}$$

   is an absolute value.
3. What goes wrong if we try to define an n-adic absolute value on $\mathbf{Q}$ for composite n, or in Example (iii) choose a reducible polynomial?
4. If a real-valued function $|\ |$ satisfies **A**.1 and **A**.3 and

$$|x + y| \leqslant 2 . \max\{|x|, |y|\},$$

show that it also satisfies **A**.2 and so is an absolute value. Given a function $|\ |$ satisfying **A**.1 and **A**.3 and

$$|x+y| \leqslant C \,.\, \max\{|x|, |y|\},$$

show that $|\ |^{\gamma}$ for suitable γ is an absolute value.
(*Hint*: Verify first that $|n| \leqslant n$ and then $|\Sigma_1^n x_i| \leqslant 2n \,.\, \max\{|x_i|\}$ by taking $n = 2^r$.)

5. Let **Q** be the rational field with the absolute value $|\ |$ and x an indeterminate. Show that an absolute value may be defined on $\mathbf{Q}(x)$ by putting $|\Sigma a_i x^i| = |\Sigma a_i e^i|$ and applying Proposition 1.1. How should the definition be modified for non-archimedean $|\ |$ to obtain a non-archimedean absolute value on $\mathbf{Q}(x)$?
6. Show that Proposition 1.2 still holds for skew fields. (*Hint*: Replace $x+y$ by $(1+yx^{-1})x$.)
7. Let $|\ |_1, |\ |_2$ be two absolute values on a field K. Show that if $|x|_2 = |x|_1$ whenever $|x|_1 > 1$ or $|x|_2 > 1$, then $|\ |_1$ and $|\ |_2$ are equal.

1.2 THE TOPOLOGY DEFINED BY AN ABSOLUTE VALUE

Just as the familiar absolute value on the field **C** of complex numbers defines **C** as a metric space, so we can use the absolute value on a field K to define a metric on K. This allows us to view K as a topological space and we can use the topological concepts to study K. But it will not be necessary to assume any knowledge of topology beyond the basic definitions.

Let K be a field with an absolute value $|\ |$ and for $x, y \in K$ define

$$d(x, y) = |x - y|. \tag{1}$$

Then d satisfies the following conditions, which the reader will have no difficulty in verifying:

M.1 $d(x, y) \geqslant 0$ *with equality if and only if* $x = y$,

M.2 $d(x, y) = d(y, x)$ *(symmetry)*,

M.3 $d(x, y) + d(y, z) \geqslant d(x, z)$ *(triangle inequality)*,

M.4 $d(x+a, y+a) = d(x, y)$ *(translation invariance)*,

M.5 $d(xa, ya) = d(x, y) \,.\, |a|$.

Thus d may be interpreted as a distance function on K and in this way K becomes a metric space in which the ring operations are continuous. The continuity of subtraction (and hence of addition) and of multiplication at the point (a, b) follows from the inequalities

$$|(x-y) - (a-b)| \leqslant |x-a| + |y-b|,$$

$$|xy - ab| = |(x-a)(y-b) + (x-a)b + a(y-b)|$$
$$\leqslant |x-a|.|y-b| + |x-a|.|b| + |a|.|y-b|.$$

In each case the right-hand side can be made arbitrarily small by taking x, y sufficiently close to a, b respectively.

To prove the continuity of inversion: $x \mapsto x^{-1}$ at $a \neq 0$, we have

$$|x^{-1} - a^{-1}| = |x|^{-1}.|x-a|.|a|^{-1}.$$

Given $\varepsilon > 0$, choose $\delta < \frac{1}{2}\min\{|a|, \varepsilon |a|^2\}$. Then for any $x \in K$ such that $|x-a| < \delta$ we have $|x| > |a| - \delta \geqslant \frac{1}{2}|a|$, hence

$$|x^{-1} - a^{-1}| \leqslant 2|a|^{-2}.|x-a| < \varepsilon,$$

and this shows inversion to be continuous at a.

We observe that the trivial absolute value defines the discrete topology: every point is open. Conversely, if $|\ |$ is a non-trivial absolute value on a field K, then there exists $a \in K$ such that $0 < |a| < 1$. For by hypothesis, $|a| \neq 1$ for some $a \neq 0$, and if $|a| > 1$, we can replace a by a^{-1}. When $0 < |a| < 1$, then we have $a \neq 0$, $a^n \to 0$ (because $|a^n| = |a|^n \to 0$), so the topology is not discrete. Two absolute values on a field K are said to be *equivalent* if they define the same topology on K. A mapping between fields with absolute values is called *analytic* if it preserves the absolute value.

Theorem 2.1

An absolute value on a field K is trivial if and only if the topology defined by it is discrete. If $|\ |_1, |\ |_2$ *are two non-trivial absolute values on K, then the following conditions are equivalent:*

(a) $|\ |_1$ *is equivalent to* $|\ |_2$*;*
(b) $|x|_1 < 1 \Rightarrow |x|_2 < 1$ *for all* $x \in K$*;*
(c) $|x|_1 > 1 \Rightarrow |x|_2 > 1$ *for all* $x \in K$*;*
(d) $|x|_1 = |x|_2^{\gamma}$ *for some positive real* γ *and all* $x \in K$.

Proof

We have just seen that the first part holds. To establish the equivalence of *(a)* ⇒ *(d)*, we clearly have *(d)* ⇒ *(a)* and to show that *(a)* ⇒ *(b)* assume that $|x|_1 < 1$. Then $x^n \to 0$ in the $|\ |_1$-metric, hence $x^n \to 0$ in the $|\ |_2$-metric and so $|x|_2 < 1$. Now *(b)* ⇒ *(c)* is clear and it only remains to show that *(c)* ⇒ *(d)*. Since $|\ |_1$ is non-trivial, there exists $a \in K$ such that $|a|_1 > 1$ and so $|a|_2 > 1$. Let b be any element of K satisfying $|b|_1 > 1$ and write

$$\lambda = \log|b|_1 / \log|a|_1$$

so that

$$|b|_1 = |a|_1^{\lambda}. \tag{2}$$

We claim that $|b|_2 \geqslant |a|_2^{\lambda}$; for if $|b|_2 < |a|_2^{\lambda}$, then

$$|b|_2 = |a|_2^{\lambda'} \quad \text{for some } \lambda' < \lambda, \tag{3}$$

hence there is a rational number $r = m/n$ such that $\lambda' < r < \lambda$ and so, by (2),

$$|b|_1 = |a|_1^{\lambda} > |a|_1^{r}.$$

Thus $|b|_1^n > |a|_1^m$, i.e. $|b^n/a^m|_1 > 1$, hence $|b^n/a^m|_2 > 1$ and so $|b|_2^n > |a|_2^m$, which gives

$$|b|_2 > |a|_2^{r} > |a|_2^{\lambda'},$$

in contradiction to (3). Hence $|b|_2 \geqslant |a|_2^{\lambda}$ and a similar argument shows that the opposite inequality holds, therefore $|b|_2 = |a|_2^{\lambda}$. We thus have

$$\frac{\log |b|_2}{\log |a|_2} = \frac{\log |b|_1}{\log |a|_1}.$$

It follows that

$$\frac{\log |b|_2}{\log |b|_1} = \frac{\log |a|_2}{\log |a|_1} = \gamma, \text{ say.}$$

Here γ depends only on a, not on b, from its definition, and we have $|b|_2 = |b|_1^{\gamma}$ for all $b \in K$ such that $|b|_1 > 1$. Replacing b by b^{-1}, we see that this also holds when $|b|_1 < 1$. If $|b|_1 = 1$, then $|ab|_1 > 1$, hence

$$|a|_2 . |b|_2 = |ab|_2 = |ab|_1^{\gamma} = |a|_1^{\gamma} = |a|_2,$$

and so $|b|_2 = 1$, therefore $|x|_2 = |x|_1^{\gamma}$ for all $x \in K$ and (d) holds.■

The following consequence is clear by using (d):

Corollary 2.2

Let K be a field with two absolute values which are equivalent. If these absolute values are non-trivial on a subfield F of K and agree on F, then they agree on K, i.e. the identity mapping is an analytic isomorphism.■

Next we turn to inequivalent absolute values and show that they satisfy a rather strong independence property, reminiscent of the Chinese remainder theorem (cf. A. 1, p. 34).

Theorem 2.3 (*Approximation theorem*)

Let $|\;|_1, \ldots, |\;|_r$ *be non-trivial absolute values on a field K which are pairwise inequivalent. Then any r-tuple over K can be simultaneously approximated;*

thus for any $\alpha_1, \ldots, \alpha_r \in K$ *and* $\varepsilon > 0$ *there exists* $\alpha \in K$ *such that*

$$|\alpha - \alpha_i|_i < \varepsilon, \quad i = 1, \ldots, r. \tag{4}$$

Proof

For any absolute value $|\ |$, if $|a| < 1$, then $a^n \to 0$, hence

$$\lim_{n \to \infty} a^n(1 + a^n)^{-1} = \begin{cases} 1 & \text{if } |a| > 1, \\ 0 & \text{if } |a| < 1. \end{cases} \tag{5}$$

(Here it is not specified what happens for $|a| = 1$.) Our first aim will be to find $c \in K$ such that

$$|c|_1 > 1, \qquad |c|_i < 1, \quad i = 2, \ldots, r. \tag{6}$$

When $r = 2$, then by Theorem 2.1 there exist $a, b \in K$ such that

$$|a|_1 > 1 \geqslant |a|_2, \qquad |b|_2 > 1 \geqslant |b|_1,$$

hence $c = ab^{-1}$ satisfies $|c|_1 > 1 > |c|_2$. This proves (6) when $r = 2$. Now take $r > 2$ and use induction; by the induction hypothesis there exists $a \in K$ such that $|a|_1 > 1 > |a|_i$ for $i = 3, \ldots, r$ and $b \in K$ such that $|b|_1 > 1 > |b|_2$. Either $|a|_2 \leqslant 1$; then we put $c_n = a^n b$ and find that $|c_n|_1 > 1 > |c_n|_2$ for all n, and when n is large enough, $|c_n|_i < 1$ holds also for $i = 3, \ldots, r$. Or $|a|_2 > 1$; then we put

$$c_n = a^n b(1 + a^n)^{-1}.$$

By (5) we have $|c_n|_1 \to |b|_1 > 1$, $|c_n|_2 \to |b|_2 < 1$, $|c_n|_i \to 0$ for $i = 3, \ldots, r$ as $n \to \infty$, so (6) holds for c_n when n is large enough.

Now take c satisfying (6); then $c^n(1 + c^n)^{-1}$ tends to 1 for $|\ |_1$ and to 0 for $|\ |_i$, $i = 2, \ldots, r$. So for any $\delta > 0$ there exists $u_i \in K$ such that $|u_i - 1|_i < \delta$, $|u_i|_j < \delta$ for $i \neq j$. Put $M = \max_i\{\Sigma_j |\alpha_j|_i\}$ and take $\delta < \varepsilon/M$; then on writing $\alpha = \Sigma\alpha_i u_i$ we have

$$|\alpha - \alpha_i|_i = |\alpha_i(u_i - 1) + \sum_{i \neq j} \alpha_j u_j|_i$$

$$\leqslant |\alpha_i(u_i - 1)|_i + \sum_{i \neq j} |\alpha_j u_j|_i \leqslant \delta M < \varepsilon,$$

and this establishes (4).■

At the root of both these theorems is the fact that inequivalent absolute values cannot be comparable. For more general valuations that are sometimes considered this need no longer hold (cf. Exercise 3 of section 1.7).

If we denote by K_i the field K with the topology induced by $|\ |_i$ and embed K in $K_1 \times \ldots \times K_r$ by the diagonal map $x \mapsto (x, x, \ldots, x)$, we can express Theorem 2.3 by saying that for inequivalent absolute values this embedding is dense.

Corollary 2.4

If $|\ |_1, \ldots, |\ |_r$ are non-trivial pairwise inequivalent absolute values on a field K, then there is no relation

$$|x|_1^{\gamma_1} \ldots |x|_r^{\gamma_r} = 1, \tag{7}$$

for fixed non-zero real numbers $\gamma_1, \ldots, \gamma_r$.

For we can choose $x \in K$ so that $|x|_1$ is small and $|x-1|_i$ is small for $i \neq 1$. Then $|x|_i^{\gamma_i}$ is close to 1 for $i > 1$, hence the left-hand side of (7) is either small (if $\gamma_1 > 0$) or large (if $\gamma_1 < 0$), and so cannot equal 1.■

In Exercise 5 (below) we shall see that we may well have a relation of type (7) when infinitely many factors are involved. This will be of importance later on.

We conclude this section with an observation due to E. Artin, that the triangle inequality is effectively a consequence of the Hausdorff separation axiom; this result is not needed later and so may be omitted.

Proposition 2.5

Let K be a field with a real-valued function $|\ |$ satisfying **A**. *1 and* **A**. *3. Then $|\ |$ defines a topology on K which is Hausdorff if and only if it is induced by an absolute value.*

Proof

The topology on K is defined by taking as a base of open neighbourhoods of $a \in K$ the sets

$$N_\varepsilon(a) = \{x \in K \mid |x-a| < \varepsilon\}.$$

If **A**. 2 holds and $a, b \in K$ are given, $a \neq b$, put $\varepsilon = \frac{1}{3} . |b-a|$; then

$$N_\varepsilon(a) \cap N_\varepsilon(b) = \varnothing,$$

because $|x-a| < \varepsilon$, $|x-b| < \varepsilon$ implies that

$$|b-a| \leqslant |b-x| + |x-a| < 2\varepsilon,$$

i.e. $3\varepsilon < 2\varepsilon$, a contradiction. Conversely, if we have a Hausdorff topology, then 0, 1 have disjoint neighbourhoods, so there exists a real $\alpha > 0$ such that no $x \in K$ satisfies $|x| < \alpha$, $|1-x| < \alpha$, thus

$$|x| + |1-x| \geqslant \alpha \quad \text{for all } x \in K. \tag{8}$$

We claim that if we replace $|\ |$ by $|\ |^r$, for suitable $r > 0$, then

$$|1-x| \leqslant 1 + |x| \quad \text{for all } x \in K. \tag{9}$$

By (2) of section 1.1 we then have an absolute value, so the topology is induced by an absolute value.

For $x = 1$, (9) is clear; if $x \neq 1$, put $y = x(x-1)^{-1}$; then (9) may be written as

$$1 \leqslant \frac{1+|x|}{|1-x|} = \frac{1}{|1-x|} + \frac{|x|}{|1-x|} = |1-y| + |y|,$$

so it remains to show that

$$|1-x| + |x| \geqslant 1 \quad \text{for all } x \in K. \tag{10}$$

Put $|x| = \xi$; then $|1-x| \geqslant \alpha - \xi$, by (8), and if we replace $|\ |$ by $|\ |^r$, the left-hand side of (8) exceeds $\xi^r + (\alpha - \xi)^r$. For varying ξ this has its least value when $\xi^{r-1} = (\alpha - \xi)^{r-1}$, i.e. $\xi = \alpha/2$, and the minimum is $2(\alpha/2)^r$. Thus

$$\xi^r + (\alpha - \xi)^r \geqslant 2(\alpha/2)^r$$

and for small enough r this is $\geqslant 1$, namely when $(\alpha/2)^r \geqslant 1/2$. With this value for r, (10) holds and the result follows.■

Exercises

1. Verify that in a field with an absolute value, the function $d(x, y)$ defined by (1) satisfies **M**.1–**M**.5.
2. Using p-adic absolute values, deduce the Chinese remainder theorem from Theorem 2.3.
3. Let K, L be topological fields, where the topologies are defined by absolute values. Show that any continuous homomorphism from K to L is a homeomorphism between K and its image in L.
4. Given r inequivalent absolute values $|\ |_1, \ldots, |\ |_r$ on a field K and real numbers $\lambda_1, \ldots, \lambda_r$ such that $|\ |_i$ assumes the value λ_i, find $\alpha \in K$ such that
$$|\alpha|_i = \lambda_i \quad (i = 1, \ldots, r).$$
5. Define a p-adic value on **Q** by $|c|_p = p^{-\nu}$, where p^ν is the exact power dividing c. Show that $\Pi |c|_i = 1$, where the product is over all absolute values, including the non-archimedean one. Obtain an analogue for the rational function field $k(x)$. (*Hint:* See section 3.2.)

1.3 COMPLETE FIELDS

Any metric space has a completion and in this section we shall discuss the completion of a field with an absolute value. This completion again has a field structure, and the complete fields with an archimedean absolute value can be classified, using standard results from analysis. At first we shall deal with arbitrary rings; this is no more difficult than the case of fields, for which the results are mainly needed later.

Let R be a ring with an absolute value $|\ |$. A sequence (c_ν) of elements of R is said to be *convergent* to $c \in R$ if $|c_\nu - c| \to 0$ as $\nu \to \infty$; we shall also write

$c_\nu \to c$ to express this fact. By a *Cauchy sequence* we understand a sequence (c_ν) such that

$$|c_\mu - c_\nu| \to 0 \quad \text{as } \mu, \nu \to \infty. \tag{1}$$

If we have a convergent sequence, say $c_\nu \to c$, then

$$|c_\mu - c_\nu| \leqslant |c_\mu - c| + |c_\nu - c|,$$

and as $\mu, \nu \to \infty$, the right-hand side tends to 0. This shows that every convergent sequence is a Cauchy sequence. If the converse holds, every Cauchy sequence is convergent, then R is said to be *complete*. The field of real numbers with the usual absolute value is complete, but not the field of rational numbers.

When the absolute valued ring R is not complete, it can be embedded in a ring $\hat{R}$ and the absolute value extended so that $\hat{R}$ is complete and R is dense in $\hat{R}$; this ring $\hat{R}$ is essentially unique and is called the *completion* of R.

Theorem 3.1
Let R be a ring with an absolute value $|\ |$. Then there exists a complete absolute valued ring $\hat{R}$ and an analytic embedding $\lambda : R \to \hat{R}$ such that the image of R is dense in $\hat{R}$, and $\hat{R}$ is unique up to an analytic isomorphism. If R is a field, then so is $\hat{R}$.

Proof
Let $R^{\mathbf{N}}$ be the direct power of $\mathbf{N}$ copies of R, i.e. the ring of all sequences (c_ν) with all operations performed componentwise. The Cauchy sequences form a subring C of $R^{\mathbf{N}}$; we claim that the set $\mathfrak{n}$ of all null sequences, i.e. sequences convergent to 0, is an ideal in C. It is clear that $\mathfrak{n}$ is closed under addition. Further, any Cauchy sequence is clearly bounded, and if $(c_\nu) \in C$, $(b_\nu) \in \mathfrak{n}$, then $|c_\nu| \leqslant M$ say, hence $|b_\nu c_\nu| \leqslant M|b_\nu| \to 0$, so $(b_\nu c_\nu) \in \mathfrak{n}$ and similarly $(c_\nu b_\nu) \in \mathfrak{n}$, which shows $\mathfrak{n}$ to be an ideal. If we identify R with the subring of constant sequences, $c \leftrightarrow (c, c, \ldots)$, then R becomes a subring of C such that $R \cap \mathfrak{n} = 0$. Therefore, if we put $\hat{R} = C/\mathfrak{n}$, the natural map

$$R \to C \to C/\mathfrak{n} = \hat{R}$$

is an embedding. To extend the absolute value from R to $\hat{R}$ we note that for any $(c_\nu) \in C$,

$$||c_\mu| - |c_\nu|| \leqslant |c_\mu - c_\nu|.$$

Hence $|c_\nu|$ is a Cauchy sequence in $\mathbf{R}$ and since the real numbers are complete, $\lim |c_\nu|$ exists. By putting

$$|(c_\nu)| = \lim |c_\nu|$$

we define an absolute value on $\hat{R}$ extending that of R; the axioms **A**.1–**A**.3 are easily checked. Clearly R is dense in $\hat{R}$: given $\alpha = (a_\nu)$, we have $a_\nu \to \alpha$, for we have $|\alpha - a_\nu| = \lim |a_n - a_\nu|$ and this tends to 0 as $\nu \to \infty$. Further, $\hat{R}$ is complete: if (α_n) is a Cauchy sequence in $\hat{R}$, and for each $n = 1, 2, \ldots$ we choose $a_n \in R$ such that $|\alpha_n - a_n| < 1/n$, then $(\alpha_n - a_n)$ is a null sequence in $\hat{R}$, hence (a_n) is a Cauchy sequence in $\hat{R}$, and so also in R. Its limit α in $\hat{R}$ satisfies $\lim |a_n - \alpha| = 0$, hence $\lim |\alpha_n - \alpha| = 0$, i.e. $\alpha_n \to \alpha$, and this shows $\hat{R}$ to be complete.

To establish the uniqueness of $\hat{R}$ we shall show that it satisfies the following universal property: given any analytic embedding $f: R \to R_1$ of R in any ring R_1 which is complete under an absolute value, there is a unique analytic embedding $f_1 : \hat{R} \to R_1$ extending f. To find f_1 take $\alpha \in \hat{R}$, say $\alpha = \lim a_n$, where (a_n) is a Cauchy sequence in R. Then $(a_n f)$ is a Cauchy sequence in R_1, so $\lim a_n f$ exists by the completeness of R_1. Moreover, if (a'_n) is another sequence tending to α, then $a_n - a'_n \to 0$, hence $a_n f - a'_n f \to 0$, and this shows $\lim a_n f$ to be independent of the sequence used; the limit may therefore be denoted by αf_1. Now it is easily verified that f_1 is an analytic embedding which extends f. It is unique because its value on a dense subring of $\hat{R}$ is prescribed. This proves the universal property of $\hat{R}$; in particular, if R is dense in R_1 then $\hat{R}$ can be embedded as a dense subring in R_1 and since $\hat{R}$ is complete, the embedding is an analytic isomorphism.

Finally, suppose that R is a field, let $c_n \to c \neq 0$ and take $p = 2/|c|$. Then $|c_n| > 1/p$ for $n > n_0$, hence $|c_n|^{-1} < p$, and so

$$|c_m^{-1} - c_n^{-1}| = |c_m^{-1}(c_n - c_m)c_n^{-1}| \leqslant p^2 . |c_n - c_m| \quad \text{for all } m, n > n_0.$$

This shows that (c_ν^{-1}) (where c_ν^{-1} is omitted if $c_\nu = 0$) is a Cauchy sequence and its limit c' clearly satisfies $cc' = c'c = 1$. Hence every non-zero element of $\hat{R}$ has an inverse and so $\hat{R}$ is a field.■

To illustrate the theorem, the completion of **Q** with the usual absolute value is **R** (but note that the completeness of **R** was used in the above proof). The completion of **Q** for the p-adic absolute value is the field $\mathbf{Q}_p$ of all p-adic numbers; these numbers can be formally represented as infinite series

$$\sum_{-N}^{\infty} a_\nu p^\nu \quad (0 \leqslant a_\nu < p)$$

and will be studied in more detail below in section 1.5. The completion of $k(x)$ for the p-adic absolute value, where $p = x - \alpha$, is the field of all formal Laurent series in $x - \alpha$; we shall also meet it later in section 1.5.

In Chapter 2 we shall study methods of extending absolute values to extension fields, and here it is best to treat the existence and uniqueness of such extensions separately. It turns out that uniqueness can be proved under quite general conditions, namely for normed finite-dimensional vector spaces over complete fields. We therefore digress briefly to establish this fact.

Let k be a field with an absolute value $|\ |$. By a *normed vector space* over k

we understand a k-space V with a real-valued function $x \mapsto \|x\|$ satisfying

N.1 $\|x\| \geqslant 0$ with equality if and only if $x = 0$,

N.2 $\|x+y\| \leqslant \|x\| + \|y\|$,

N.3 $\|\alpha x\| = |\alpha| . \|x\|$, for all $x, y \in V, \alpha \in k$.

For example, any field extension of k, with an absolute value extending that of k, is a normed vector space.

Any finite-dimensional k-space V has at least one norm, the *cubical norm*, defined as follows. Take any basis $e_1, \ldots, e_n$ of V and for $x = \Sigma \alpha_i e_i$ put $\|x\| = \max_i \{|\alpha_i|\}$; it is not hard to verify that this is in fact a norm. This norm still depends on the choice of basis, but as we shall see below, the topology induced is independent of this choice. In any normed space we can define completeness as in the case of fields, using the norm in place of the absolute value. A point of interest to us is that in the finite-dimensional case the completeness of V follows from that of k:

Lemma 3.2

Let V be a finite-dimensional vector space over a complete absolute valued field k. Then V is complete in the topology induced by the cubical norm, and any other norm induces the same topology. In particular, if k is discrete, then so is V.

Proof

To prove that V is complete in the cubical norm, let $x_\nu = \Sigma \xi_{i\nu} e_i$ be a Cauchy sequence. Then $x_\mu - x_\nu \to 0$ and so $|\xi_{i\mu} - \xi_{i\nu}| \to 0$ as $\mu, \nu \to \infty$, for $i = 1, \ldots, n$. Hence $\xi_{i\nu}$ converges to ξ_i say, by the completeness of k. Write $x = \Sigma \xi_i e_i$; then

$$\|x - x_\nu\| = \max_i \{|\xi_i - \xi_{i\nu}|\} \to 0,$$

hence $x_\nu \to x$ and this shows V to be complete.

If $N(x)$ is any other norm, we have to show that this defines the same topology as the cubical norm $\|.\|$. Given $x = \Sigma \xi_i e_i$, we have

$$N(x) \leqslant \max_i |\xi_i| . \Sigma N(e_i) = \|x\| . M$$

for some M independent of x, so the cubical norm is finer than the N-topology (i.e. $\|x_\nu\| \to 0$ implies $N(x_\nu) \to 0$). Conversely, suppose that (x_ν) is a sequence such that $N(x_\nu) \to 0$ but $\|x_\nu\| \nrightarrow 0$. Write $x_\nu = \Sigma \xi_{i\nu} e_i$; if $|\xi_{i\nu}| \to 0$ for $i = 1, \ldots, n$, then

$$\|x_\nu\| = \max_i |\xi_{i\nu}| \to 0,$$

which is false, so for some i, say $i = 1$, $|\xi_{1\nu}| \nrightarrow 0$. By passing to a subsequence we may assume that $|\xi_{1\nu}| \geqslant \varepsilon$ for some $\varepsilon > 0$ and all ν. Put

$$y_\nu = \xi_{1\nu}^{-1} x_\nu = \Sigma \eta_{i\nu} e_i;$$

then $\eta_{1\nu} = 1$ by construction and $N(y_\nu) \leqslant \varepsilon^{-1} N(x_\nu)$, hence $N(y_\nu) \to 0$, thus $\Sigma_2^n \eta_{i\nu} e_i \to -e_1$ in the N-topology. Let W be the subspace spanned by $e_2, \ldots, e_n$; it is $(n-1)$-dimensional and so by the induction hypothesis has a unique topology and is complete, hence closed. Therefore W contains e_1, which is a contradiction; it follows that $\|x_\nu\| \to 0$ and the N-topology is just the cubical topology. The last assertion is clear since k is complete under the trivial norm.■

This result shows in particular that the cubical norm is essentially independent of the choice of basis. It also shows the uniqueness of the extensions of absolute values:

Theorem 3.3.
Let K be a complete absolute valued field. Then any algebraic extension E of K has at most one extension of the absolute value and it is complete for the induced topology.

Proof
Let $|\ |_1$, $|\ |_2$ be two absolute values on E extending the given absolute value $|\ |$ on K; we have to show that $|\alpha|_1 = |\alpha|_2$ for all $\alpha \in E$, so on replacing E by $K(\alpha)$ we may take E to be of finite degree over K. By Lemma 3.2 both absolute values determine the same topology and E is complete in this topology. If either $|\ |_1$ or $|\ |_2$ is trivial, then the topology is discrete and both must be trivial. Otherwise the extension is unique by Corollary 2.2.■

The archimedean absolute values can easily all be determined. This was first accomplished by Ostrowski in 1918; his proofs were simplified by E. Artin, and we shall follow the latter source below.

Theorem 3.4 (*Ostrowski's first theorem*)
Any archimedean absolute value on $\mathbf{Q}$ is equivalent to the usual absolute value.

Proof
Let ϕ be any archimedean absolute value on $\mathbf{Q}$; we have to show that $\phi(a) = |a|^\gamma$ for some γ. We take any integers $r, s > 1$ and express s in the base of r:

$$s = a_0 + a_1 r + a_2 r^2 + \ldots + a_\nu r^\nu, \quad \text{where } 0 \leqslant a_i < r,\ a_\nu \neq 0. \tag{2}$$

It follows that $s \geqslant r^\nu$, hence

$$\nu \leqslant \log s / \log r. \tag{3}$$

By (2) we have

$$\phi(s) \leqslant \phi(a_0) + \phi(a_1)\phi(r) + \ldots + \phi(a_\nu)\phi(r)^\nu$$

and it is clear that $\phi(a_i) \leqslant a_i < r$, therefore

$$\phi(s) \leqslant r[1 + \phi(r) + \ldots + \phi(r)^\nu].$$

Each term in the bracket is at most 1 if $\phi(r) \leqslant 1$ and at most $\phi(r)^\nu$ otherwise, hence

$$\phi(s) \leqslant r(1 + \nu) \max\{1, \phi(r)^\nu\}.$$

By (3) this can be written

$$\phi(s) \leqslant r\left(1 + \frac{\log s}{\log r}\right) \max\{1, \phi(r)^{\log s/\log r}\}.$$

We now replace s by s^n and take nth roots:

$$\phi(s) \leqslant r^{1/n}\left(1 + n\frac{\log s}{\log r}\right)^{1/n} \max\{1, \phi(r)^{\log s/\log r}\}.$$

If we now let $n \to \infty$, we obtain by (7) of section 1.1,

$$\phi(s) \leqslant \max\{1, \phi(r)^{\log s/\log r}\}. \tag{4}$$

Since ϕ is archimedean, we have $\phi(n_0) > 1$ for all n_0 by Proposition 1.2; so we can rewrite (4) as

$$\phi(s) \leqslant \phi(r)^{\log s/\log r},$$

or on taking logarithms,

$$\frac{\log \phi(s)}{\log s} \leqslant \frac{\log \phi(r)}{\log r}.$$

By symmetry we have equality, say both sides are equal to γ. Thus

$$\log \phi(r) = \gamma \,.\, \log r \quad \text{for all } r > 1,$$

and so $\phi(r) = r^\gamma$. Now

$$\phi(-r) = \phi(r) = r^\gamma \quad \text{and} \quad \phi(r/s) = (r/s)^\gamma,$$

hence $\phi(a) = |a|^\gamma$ for all $a \in \mathbf{Q}$, as claimed.■

In section 1.4 we shall determine all absolute values on **Q** and find that the only non-archimedean absolute values up to equivalence are the p-adic absolute values, for the various primes p.

It is possible to determine all fields which are complete under an archimedean absolute value: the only cases are **R** and **C**. As this result is not needed later we shall only present an outline of the proof. It is based on the Gelfand–Mazur theorem, which states that a field K containing **C** which as a **C**-space is a complete normed space (i.e. a complex Banach algebra) must be equal to **C** (cf. Rudin (1966), p. 355).

Theorem 3.5 (*Ostrowski's second theorem*)
Let K be a field with an archimedean absolute value for which it is complete. Then K is analytically isomorphic either to **R** *or to* **C**.

Outline proof.
Corollary 1.3 shows that K is of characteristic 0, so its prime subfield is **Q**, and by Theorem 3.4, the restriction of the absolute value to **Q** is equivalent to the usual absolute value. By replacing the absolute value on K by a suitable power we ensure that it equals the absolute value on **Q**, but now a separate verification of the triangle inequality (which might be violated when we pass to a power) is necessary (cf. e.g. A.2, p. 278 or also Exercise 3 below).

Since K is complete, it contains **R** as the completion of **Q**. We distinguish two cases:

(i) The equation $x^2 + 1 = 0$ is irreducible over K. Then we adjoin a root i of this equation to K and on $K(i)$ define a norm by

$$\| x + yi \| = (|x|^2 + |y|^2)^{1/2} \quad \text{for } x, y \in K.$$

Then $K(i)$ is a normed K-space, complete by Lemma 3.2 and so, by the Gelfand–Mazur theorem, $K(i) = \mathbf{C}$. Since $\mathbf{R} \subseteq K \subset \mathbf{C}$, it follows that $K = \mathbf{R}$.

(ii) K contains i satisfying $i^2 = -1$. Then $K \supseteq \mathbf{C}$ and the same argument as before shows that $K = \mathbf{C}$. ■

We remark that locally compact fields are complete. Since we have defined completeness in terms of absolute values, we shall prove this fact only for absolute valued fields, although it holds quite generally. In any case the result will not be needed later and so can be omitted without loss of continuity.

Proposition 3.6
Let K be a field with an absolute value. If in the induced topology K is locally compact, then it is complete.

Proof
For discrete fields this is clear, so we may assume that the absolute value is non-trivial. Let $N = \{x \in K \mid |x| \leqslant \gamma\}$ be a compact neighbourhood of zero and take any Cauchy sequence (c_ν). This sequence is bounded, say $|c_\nu| \leqslant M$; if $a \in K$ is such that $0 < |a| < 1$, then for sufficiently large n, $|a|^n M \leqslant \gamma$, and with this value of n, $|a^n c_\nu| \leqslant \gamma$ for all ν. Thus $a^n c_\nu \in N$, and by compactness we can find a convergent subsequence, say $a^n c_{\nu_i} \to p$. Since (c_ν) is a Cauchy sequence, so is $(a^n c_\nu)$, hence $a^n c_\nu \to p$ and so $c_\nu \to a^{-n} p$. Thus every

Cauchy sequence is convergent and K is complete.■

In general a complete space need not be locally compact; in fact a linear space (or more generally, a uniform space) is compact if and only if it is complete and precompact (= totally bounded, cf. Bourbaki (1966)). Thus a complete field is locally compact if and only if it has a precompact neighbourhood of zero. But in fact every field which is complete under an archimedean absolute value is locally compact, by Theorem 3.5. In Proposition 4.6 below we shall obtain a condition in the non-archimedean case.

Exercises

1. Show that for any absolute valued ring R with completion $\hat{R}$, any continuous homomorphism of R into a complete ring I extends to a continuous homomorphism of $\hat{R}$ into I.
2. Let V be a vector space over an absolute valued field k with a function $\phi(x)$ satisfying **N**.1 and **N**.2 and $\phi(\alpha x) \leqslant |\alpha| . \phi(x)$ for all $\alpha \in k$, $x \in V$. Show that ϕ satisfies **N**.3 (and hence is a norm on V).
3. Let ϕ be an absolute value on a field K and $\gamma > 1$, and suppose that the function ψ defined by $\psi(x) = \phi(x)^\gamma$ agrees with the usual absolute value on $\mathbf{Q}$. Show that
$$\psi(x+y) \leqslant 2\max\{\psi(x), \psi(y)\}$$
and deduce that ψ is again an absolute value.
4. Let K be a skew field with an archimedean absolute value for which it is complete. Show that if $x^2 + 1 = 0$ has a root in the centre of K, then $K \cong \mathbf{C}$. Show further that if $x^2 + 1 = 0$ has a root i in K which is not in the centre, then K is analytically isomorphic to the quaternions. (*Hint*: Verify that the centre of K is closed in K, hence complete.)
5. Let A be a dense subring of a complete absolute valued ring R. Show that any analytic homomorphism from A to a complete absolute valued ring S can be extended in just one way to an analytic homomorphism of R into S.
6. Use (4) to give another proof that $(c) \Rightarrow (b)$ in Proposition 1.2.

1.4 VALUATIONS, VALUATION RINGS AND PLACES.

In section 1.3 we obtained a fairly complete picture of fields with an archimedean absolute value. We now turn to consider the non-archimedean case in more detail. As Krull (1931) has observed, a non-archimedean absolute value uses only the multiplicativity and the ordering of the real numbers, since the addition of values (which occurred in **A**.2) is no longer needed. This means that we can take the values to lie in any ordered multiplicative group, but for the moment we shall confine ourselves to the case of the real numbers. It will be convenient to work with additive groups instead; formally we can make the transition by taking logarithms (to any fixed base > 1). Thus if K is a field with a non-archimedean absolute value | | and we define

$$v(x) = -\log|x| \quad (x \in K),$$

then v is a function on K with the following properties:

V.1 $v(x)$ is a real number for any $x \neq 0$, while $v(0) = \infty$,

V.2 $v(x+y) \geqslant \min\{v(x), v(y)\}$,

V.3 $v(xy) = v(x) + v(y)$.

Such a function is called a *valuation* on K; more precisely we speak of a *real-valued* or *rank* 1 valuation. We note that by Theorem 2.1, two valuations v_1, v_2 are equivalent if and only if there is a real positive constant γ such that $v_1(x) = \gamma v_2(x)$ for all $x \in K$. As before, we have the *trivial* valuation, defined by

$$v(x) = \begin{cases} 0 & \text{if } x \neq 0, \\ \infty & \text{if } x = 0. \end{cases} \tag{1}$$

From **V**.1–**V**.3 it is clear that for any valuation the values taken on K form a subgroup Γ of the additive group of real numbers. Here two cases can arise, besides the trivial valuation, which corresponds to the subgroup consisting of 0 alone. Either Γ has a least positive element λ; in that case $\Gamma = \lambda\mathbf{Z}$. For clearly Γ contains any multiple of λ; now if $\alpha \in \Gamma$ we can find $n \in \mathbf{Z}$ such that $0 \leqslant \alpha - n\lambda < \lambda$, and by the definition of λ we must have $\alpha - n\lambda = 0$, so $\alpha = n\lambda$. Thus Γ is infinite cyclic; the valuation in this case is said to be *discrete of rank* 1, or also *principal*. Or Γ has no least positive element; in that case Γ is clearly dense in $\mathbf{R}$, in the sense that between any two real numbers there is an element of Γ. Most of our valuations are in fact principal; the main exceptions occur in sections 1.6 and 1.7 and Chapter 5. For a principal valuation we can always find an equivalent one with precise value group $\mathbf{Z}$; such a valuation is said to be *normalized*.

It is a natural step to extend the above definition of a valuation by allowing as value group any totally ordered abelian group. Even though we shall mainly be concerned with principal valuations, it will help our understanding to consider the general situation, and we shall briefly do so in this section and in sections 1.6 and 1.7.

Thus Γ will be an abelian group, written additively, with a total ordering preserved by the group operation, i.e.

$$\alpha \leqslant \beta \Rightarrow \alpha + \gamma \leqslant \beta + \gamma.$$

We adjoin to Γ a symbol ∞ with the rules

$$\alpha + \infty = \infty, \qquad \alpha < \infty \quad \text{for all } \alpha \in \Gamma.$$

By a *general* or *Krull valuation* on a field K with values in Γ, or a *valuated field*, we shall understand a function v from K to $\Gamma \cup \{\infty\}$ such that $v(x) = \infty$

if and only if $x=0$, and **V**.2 and **V**.3 hold. If the image of $K^\times$ under v is the whole of Γ we call Γ the *precise* value group. In section 1.7 we shall meet conditions under which general valuations reduce to the real case.

As for non-archimedean absolute values, we have the following consequence of **V**.2 for any general valuation:

$$v(x+y)=\min\{v(x), v(y)\}, \quad \text{whenever } v(x)\neq v(y). \tag{2}$$

Further, we have the *principle of domination*:

Given $x_1, \ldots, x_n$ in a valuated field, if $x_1+\ldots+x_n=0$, then at least two of the $v(x_i)$ are equal.

For, from **V**.2, we obtain by induction, $v(a_1+\ldots+a_r)\geqslant \min_i\{v(a_i)\}$. If the $v(x_i)$ are all different, we can renumber the x's so that

$$v(x_1)<v(x_2)< \ldots <v(x_n).$$

Then $-x_1=x_2+\ldots+x_n$ and

$$v(x_1)=v(-x_1)<\min\{v(x_2), \ldots, v(x_n)\},$$

which is a contradiction.

Given any general valuation v on a field K, we define

$$V=\{x\in K \mid v(x)\geqslant 0\}. \tag{3}$$

It is easily verified that V is a subring of K; moreover, since the value group is totally ordered, V has the property that

$$\text{for any } x\in K^\times \quad \text{either} \quad x\in V \text{ or } x^{-1}\in V. \tag{4}$$

If we write $V^{-1}=\{x\in K^\times \mid x^{-1}\in V\}$, then (4) may be expressed briefly as $V\cup V^{-1}=K$. A subring V of K satisfying (4) is called a *valuation ring*; its elements are the *valuation integers* or *v-integers*.

We see that with every valuation on K there is associated a valuation ring in K and we shall find that conversely, every valuation ring on K arises from a valuation in this way. For the moment we note that $V=K$ precisely when v is the trivial valuation.

In any valuation ring V the collection of principal ideals $\{aV\}$ is totally ordered by inclusion, for $aV\subseteq bV$ holds precisely if $ab^{-1}\in V$. It follows that there is a unique maximal ideal $\mathfrak{p}$ (not necessarily principal), consisting of all the non-units in V. When V is defined in terms of a valuation as in (3), then

$$\mathfrak{p}=\{x\in K \mid v(x)>0\}.$$

The residue class ring $V/\mathfrak{p}$ is a field, because every element not in $\mathfrak{p}$ is a unit; this field $V/\mathfrak{p}$ is called the *residue class field* of V.

Now let K be any field containing a valuation ring V and denote by $U=V\cap V^{-1}$ the group of units in V. In $K^\times$ we have the relation of divisibility relative to V: $a \mid b$ if and only if $ba^{-1} \in V$. This is a preordering of K:

1. $a \mid a$ (reflexive);

2. $a \mid b,\ b \mid c \Rightarrow a \mid c$ (transitive);

3. $a \mid b \Rightarrow ac \mid bc$ (multiplicative).

If $a \mid b$ and $b \mid a$, we say that a and b are *associated*. If associated elements are equal, we have an ordering, but this will not usually be the case. In general the classes of associated elements are just the cosets of U in $K^{\times}$. Let us write $\Gamma = K^{\times}/U$ and denote the natural homomorphism from $K^{\times}$ to Γ by v; we can order Γ by writing $v(a) \geqslant v(b)$ if and only if $b \mid a$. With this definition Γ is a totally ordered group and v is a valuation on K (writing $v(0) = \infty$ as before). For **V**.1 (with Γ replacing **R**) and **V**.3 hold by definition; to prove **V**.2 we note that this clearly holds if $x = 0$ or $y = 0$. If $x, y \neq 0$, assume that $v(x) \geqslant v(y)$, say; then $y \mid x$, i.e. $xy^{-1} \in V$. Hence $xy^{-1} + 1 \in V$ and so $v(x+y) \geqslant v(y)$, which proves **V**.2. Moreover, the valuation ring associated with v is easily seen to be V.

In the other direction, if v is a valuation on K with associated valuation ring V and the valuation defined by V is w, then it is not hard to check that there is an order-preserving isomorphism α between the value groups of v and w such that

$$w(x) = v(x)^{\alpha}. \tag{5}$$

Let us call two general valuations related in this way *equivalent*. For real valuations this reduces to the notion introduced in section 1.2, because any order-preserving automorphism of the additive group of **R** has the form $c^{\alpha} = kc$ for some positive real k (cf. Exercise 2 and section 1.6). We can now sum up our conclusion as follows.

Theorem 4.1

Let K be any field. Then the correspondence $v \leftrightarrow V$ defined by equation (3) is a natural bijection between valuation rings on K and equivalence classes of general valuations on K. ■

Let us look in more detail at the case of principal valuations, which is important for us later. By a *principal valuation ring* we understand a valuation ring which is also a principal ideal domain. Let V be a principal valuation ring in a field K; its unique maximal ideal has the form pV, where p is clearly irreducible (i.e. a non-unit without proper factors). We claim that every element c of K has the unique form

$$c = p^r u, \text{ where } r \in \mathbf{Z} \text{ and } u \text{ is a unit in } V. \tag{6}$$

In the first place we have $\cap\, p^n V = 0$. For the left-hand side is an ideal in V, hence of the form qV. Since $pV \supset p^2 V \supset \ldots$, we have

$$qpV = \cap p^{n+1}V = qV,$$

so $q = qpa$ for some $a \in V$. But $1 - pa$ is a unit in V, because it is not in pV, and $q(1 - pa) = q - qpa = 0$, hence $q = 0$, as claimed.

Given $c \in K^{\times}$, suppose first that $c \in V$. Since $c \neq 0$, we have $c \notin p^rV$ for some r, hence the set of numbers n such that $c \in p^nV$ is bounded above. If r is the largest value and $c = p^ru$, then u must be a unit in V, because otherwise $c \in p^{r+1}V$. Thus we have (6) in this case. The expression (6) is unique, for if $c = p^ru = p^sv$, where u, v are units and $r \leqslant s$ say, then $u = p^{s-r}v$, hence $r = s$ and $u = v$. There remains the case where $c \notin V$; then $c^{-1} \in V$, so we obtain an expression (6) for c^{-1}, say $c^{-1} = p^sv$, hence we find that $c = p^{-s}v^{-1}$, which again has the form (6).

An element p generating the maximal ideal of a principal valuation ring is called a *prime element* or a *uniformizer*. It is clear that the valuation defined by a principal valuation ring is principal: if c is given by (6), then we have $v(c) = r$. Conversely, every principal valuation clearly defines a principal valuation ring V; its ideals are of the form p^nV, where p is a uniformizer. Thus in the principal case Theorem 4.1 takes the following form:

Theorem 4.2

On any field K, the correspondence $v \leftrightarrow V$ defined by Theorem 4.1 is a natural bijection between (equivalence classes of) principal valuations and principal valuation rings. The normalized valuation corresponding to V is given by $v(c) = r$ if $c = p^ru$, where p is any uniformizer and u is a unit. ■

We remark that the normalized valuation is independent of the choice of uniformizer, while the uniformizer is determined up to a unit factor.

Let us look again at some of the examples in section 1.1. The p-adic absolute value clearly corresponds to a principal valuation, with valuation ring consisting of all rational fractions with denominator prime to p. If we put

$$c = p^{\nu}\,.\,\frac{m}{n}, \quad \text{where } m, n, \nu \in \mathbf{Z}, n > 0, p \nmid mn,$$

then $v(c) = \nu$ is the corresponding normalized valuation. The residue class field is $\mathbf{F_p}$, the field of p elements.

Next consider the absolute value defined by an irreducible polynomial $p(x)$ over a field k. This leads again to a principal valuation $v(\phi) = \nu$ if $\phi = p^{\nu}\,.\,f/g$, where $p \nmid fg$. This valuation is trivial on k; we also say that v is a valuation of $k(x)$ *over* k. The residue class field in this case is the finite extension of k defined by a root of $p(x) = 0$. For it is generated by the image $\bar{x}$ of x in the quotient $V/(p(x))$, hence $p(\bar{x}) = 0$. In particular, for an algebraically closed field like $\mathbf{C}$, all irreducible polynomials are linear and the residue class field is then just $\mathbf{C}$ itself.

We can now determine all the non-archimedean absolute values on $\mathbf{Q}$; by the earlier remarks it comes to the same to find all real valuations; more generally we shall determine all general valuations on **Q**.

Proposition 4.3
*Any general valuation on **Q** is either trivial or equivalent to a p-adic valuation, for some prime number p.*

Proof
By Theorem 4.1 it is enough to determine all valuation rings in **Q**. Any valuation ring V contains 1 and hence all of **Z**. Let $\mathfrak{p}$ be the maximal ideal in V and consider $\mathfrak{p} \cap \mathbf{Z}$; this is a prime ideal in **Z**. Either $\mathfrak{p} \cap \mathbf{Z} = 0$; then $V = \mathbf{Q}$ and we have the trivial valuation. Or $\mathfrak{p} \cap \mathbf{Z} = p\mathbf{Z}$ for a prime number p; then any integer n is either divisible by p or prime to p, so either $n \in p\mathbf{Z}$ or n is a unit in V. It follows that V consists of all fractions with denominator prime to p, and this corresponds to the p-adic valuation.■

In a similar way we can determine all valuations of $k(x)$ over k; they are associated with the irreducible polynomials, together with an extra one, defined by x^{-1}.

Proposition 4.4
Let k be any field. Any general valuation on the rational function field $k(x)$ over k which is non-trivial is either associated to an irreducible polynomial over k, or to x^{-1}. In particular, any valuation on $k(x)$ over k is either principal or trivial.

Proof
Let V be a valuation ring in $k(x)$; then $V \supseteq k$, because the valuation is over k. Suppose first that $x \in V$, so that $V \supseteq k[x]$. We again consider $\mathfrak{p} \cap k[x]$, where $\mathfrak{p}$ is the maximal ideal of V. If $\mathfrak{p} \cap k[x] = 0$, then $V = k(x)$ and the valuation is trivial. Otherwise $\mathfrak{p} \cap k[x]$ is a non-zero prime ideal of $k[x]$, hence of the form $pk[x]$, where p is an irreducible polynomial, and now the same argument as before shows that we have the valuation associated to p. This depends on the fact that $k[x]$ is a principal ideal domain, so any polynomial not divisible by p is a unit mod p.

If $x \notin V$, then $x^{-1} \in V$, in fact $x^{-1} \in \mathfrak{p}$ and if we apply the same argument to $y = x^{-1}$ we find the valuation associated to y. Let

$$f = a_0 x^n + \ldots + a_n, \quad a_0 \neq 0;$$

in terms of y we have

$$f = a_0 y^{-n} + \ldots + a_n = y^{-n}(a_0 + a_1 y + \ldots + a_n y^n).$$

Here the second factor is a unit and y is a uniformizer, therefore $v(f) = -n = -\deg f$, where deg refers to the degree in x. Thus for $\phi = f/g$ we have $v(\phi) = \deg g - \deg f$.■

Consider the case $k = \mathbf{C}$; here the irreducible polynomials are $x - \alpha(\alpha \in \mathbf{C})$, so every complex number defines a valuation on $\mathbf{C}(x)$; in addition we have a valuation corresponding to x^{-1}. These valuations just correspond to the points on the Riemann sphere, and the representation $\phi = (x - \alpha)^{\nu} f/g$ or $x^{-\nu} f/g$ may be regarded as indicating the leading term of ϕ at the point $x = \alpha$ or $x = \infty$.

The residue class field leads to yet another way of viewing valuations. Let K, F be two fields; by a *place* of K in F we understand a map $\phi : K \to F \cup \{\infty\}$ such that ϕ restricted to $F\phi^{-1}$ is a ring homomorphism and $x\phi = \infty \Rightarrow x \neq 0$ and $(x^{-1})\phi = 0$. Two places ϕ_1, ϕ_2 of K in F_1, F_2 are said to be *isomorphic* if $F_1\phi_1^{-1} = F_2\phi_2^{-1}$ and there is an isomorphism $\theta : F_1 \to F_2$ such that the diagram below commutes. If k is a subfield of K on which ϕ is an isomorphism, we shall call ϕ a place *over* k.

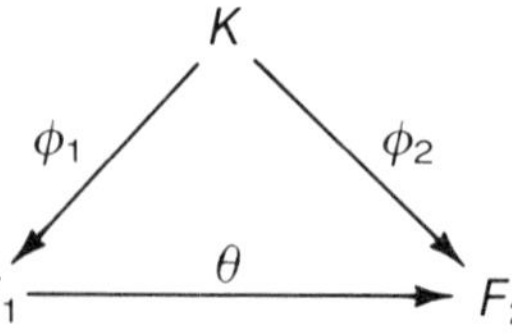

For example, consider the rational function field $k(x)$. Given $\alpha \in k$, we can write any $u \in k(x)$ in the form $u = f/g$, where $f, g \in k[x]$ and $f(\alpha)$, $g(\alpha)$ are not both zero. We define the *value* of u at α as

$$u\phi = \begin{cases} u(\alpha) & \text{if } u = f/g, \text{ where } g(\alpha) \neq 0, \\ \infty & \text{if } u = f/g, \text{ where } f(\alpha) \neq 0, g(\alpha) = 0. \end{cases}$$

In this way we have a place of $k(x)$ in k (and over k); of course this just corresponds to the residue class field of the valuation defined by $x - \alpha$ as uniformizer. This connexion is quite general, as the next result shows.

Theorem 4.5

There is a natural bijection between the isomorphism classes of places on a field K and the valuation rings in K. Moreover, in this bijection the places over a subfield k of K correspond to the valuation rings containing k.

Proof

Let ϕ be a place of K in F: then $V = F\phi^{-1}$ is a valuation ring in K, for it clearly is a subring of K, and for any $x \in K^{\times}$ we have

$$x \notin V \Leftrightarrow x\phi = \infty \Leftrightarrow x^{-1}\phi = 0 \Leftrightarrow x^{-1} \text{ is in } V \text{ and is a non-unit.}$$

Conversely, given a valuation ring V in K, with maximal ideal $\mathfrak{p}$ and residue class field $F = V/\mathfrak{p}$, let us write $x \mapsto \bar{x}$ for the natural homomorphism and define $\phi : K \to F \cup \{\infty\}$ by

$$x\phi = \begin{cases} \bar{x} & \text{if } x \in V, \\ \infty & \text{if } x \notin V. \end{cases}$$

Then ϕ is easily verified to be a place of K in F. Further, two places correspond to the same valuation ring if and only if they are isomorphic, and it is clear that a place is over k precisely when the corresponding valuation ring contains k.■

We now see that for any field K there is a triple correspondence ('trijection') between the following sets: valuation rings in K, equivalence classes of valuations on K, isomorphism classes of places on K. Whereas a valuation is determined only up to equivalence and a place up to isomorphism, a valuation ring is completely determined. This often makes it easier to study general valuations through their valuation rings. However, for principal valuations there is the uniquely determined normalized form, which is also easily handled.

The method of completion given in section 1.3 clearly applies to any field with a real-valued valuation, since the resulting topology can also be described by an absolute value. For more general valuations this method does not apply; there is a process of forming completions in this case, but this will not be needed here and we refer to Schilling (1950) for the details.

We go on to consider how the residue class field and the value group change on completion. Let K be a field with a real-valued valuation v and $\hat{K}$ its completion; write V, W for the valuation rings in K, $\hat{K}$ and denote their maximal ideals by $\mathfrak{p}$, $\mathfrak{P}$ respectively. We have

$$\mathfrak{P} \cap V = \mathfrak{p}, \qquad V + \mathfrak{P} = W. \tag{7}$$

The first equation is clear; for the second we note that any element $c \in W$ is the limit of a sequence in V, so there exists $c_1 \in V$ such that $c - c_1 \in \mathfrak{P}$, and now the equations (7) follow. As a consequence we have

$$V/\mathfrak{p} = V/(V \cap \mathfrak{P}) \cong (V + \mathfrak{P})/\mathfrak{P} \cong W/\mathfrak{P}.$$

This shows the residue class field of the completion $\hat{K}$ to be canonically isomorphic to that of K. The value group of the valuation on $\hat{K}$ is evidently the completion (with respect to the ordering) of that on K. In the case of main concern for us, of a field with a principal valuation, the value group is $\mathbf{Z}$ and remains unchanged by completion.

Let us return to the case of a field K with a principal valuation v. If V is the corresponding valuation ring and pV the maximal ideal, then the powers p^nV form a base for the neighbourhoods of zero and as we have seen, $\cap p^nV = 0$. Since p^nV is open and a subgroup of the additive group of K, all its cosets are open; hence p^nV as complement of the union of the remaining cosets is also closed. Thus 0 has a base of neighbourhoods which are open and closed, so K is totally disconnected. If U denotes the group of units in V, we have

$$K^{\times} \cong \langle p \rangle \times U, \tag{8}$$

where $\langle p \rangle$ denotes the subgroup generated by the uniformizer p. Let us write

$$U_n = 1 + p^n V = \{x \in K \mid v(x-1) \geqslant n\} \quad \text{for } n \geqslant 1;$$

in particular, U_1 is the set of units that are $\equiv 1 \pmod p$, sometimes called 1-*units* (*Einseinheiten*). Writing $U_0 = U$, we have the chain of subgroups

$$K^{\times} \supset U = U_0 \supset U_1 \supset U_2 \ldots, \qquad \cap U_n = \{1\}; \tag{9}$$

$$U_0/U_1 \cong F^{\times}, \qquad U_n/U_{n+1} \cong F^{+} \quad (n \geqslant 1), \tag{10}$$

where F denotes the residue class field. The first relation in (10) follows because the natural homomorphism $V \to V/p \cong F$ defines a homomorphism of $U = U_0$ on to $F^{\times}$ with kernel U_1. If $n \geqslant 1$, any element of U_n has the form $1 + p^n a$, $a \in V$, and this lies in U_{n+1} precisely if $a \in pV$; moreover, we have

$$(1 + ap^n)(1 + bp^n) = 1 + (a + b + abp^n)p^n,$$

and this proves the rest of (10).

We can now also describe the complete fields with a non-archimedean absolute value (or equivalently, a valuation):

Propositions 4.6

Let K be a valuated field which is complete. Then either K has prime characteristic, or it is a $\mathbf{Q}_p$-algebra, where $\mathbf{Q}_p$ is the field of p-adic numbers, for some prime p.

Proof

Suppose that char $K = 0$ and let E be the closure of the prime subfield $\mathbf{Q}$ of K. By Proposition 4.3, the valuation restricted to $\mathbf{Q}$ is the p-adic valuation for some prime p, hence E is the completion $\mathbf{Q}_p$ of $\mathbf{Q}$ and so K is a $\mathbf{Q}_p$-algebra.■

Here a more precise result (as in Theorem 3.5) is not to be expected, because $\mathbf{Q}_p$ is of infinite degree under its algebraic closure (in contrast to $\mathbf{R}$). If in Proposition 4.6, K is assumed to be locally compact, then considerations of volume (using the Haar integral) show that K must be finite-dimensional over $\mathbf{Q}_p$; this applies even when K is a skew field (cf. Weil, (1967), p. 11).

Exercises

1. Let R be an integral domain with field of fractions K. Defining divisibility and associated elements as in the text, show that the classes of associated elements form a partially ordered abelian group, under the ordering induced by divisibility, with the property:

$$v(x), v(y) \geqslant \gamma \Rightarrow v(x+y) \geqslant \gamma.$$

Show that conversely, any such function v on K defines a ring (by means of (3)) with K as its field of fractions, provided that the group of values is directed (i.e. for any α, β there exists $\gamma \geqslant \alpha, \gamma \geqslant \beta$).

2. Show that any order-preserving automorphism of $\mathbf{R}^+$ is of the form $x \mapsto \lambda x$, for some real positive λ. (*Hint:* See the proof of Corollary 6.4.)
3. Show that a valuated field is locally compact if and only if it is complete, with a principal valuation and a finite residue class field.
4. Let k be a perfect field, k_a its algebraic closure and denote by G the Galois group of k_a/k. Show that the isomorphism classes of places of $k(x)$ over k are in a natural bijection with the orbits of G in its action on k_a, together with a point at ∞.
5. Prove an analogue of Proposition 1.1 for general valuations.
6. Show that a rational number is a square if and only if it is a square in $\mathbf{R}$ and in $\mathbf{Q}_p$ for each prime p.

1.5 THE REPRESENTATION BY POWER SERIES

As we have seen in section 1.3, every field with an absolute value has a completion, and this certainly applies to fields with a rank 1 valuation. We shall now examine this completion more closely in the case of a principal valuation and show how its elements can be represented by power series.

Let K be any field with a principal valuation v and uniformizer π. Taking v to be normalized, we have $v(\pi) = 1$. We denote by V the valuation ring; then (π) is the maximal ideal and $V/(\pi)$ the residue class field. Let us take a transversal A of (π) in V; thus each element of V is congruent (mod π) to exactly one element of A. For convenience we assume that $0 \in A$, so that 0 is represented by itself.

Given $c \in K^{\times}$, suppose that $v(c) = r$. Then $c\pi^{-r}$ is a unit, so there exists a unique $a_r \in A$ such that $c\pi^{-r} \equiv a_r \pmod{\pi}$; moreover, $a_r \neq 0$. It follows that

$$v(c - a_r\pi^r) \geqslant r + 1,$$

so there exists a unique $a_{r+1} \in A$ such that

$$(c - a_r\pi^r)\pi^{-r-1} \equiv a_{r+1} \pmod{\pi}.$$

Now an induction shows that for any $n \geqslant r$ we have a unique expression

$$c \equiv a_r\pi^r + a_{r+1}\pi^{r+1} + \ldots + a_n\pi^n \pmod{\pi^{n+1}}, \text{ where } a_i \in A,\ a_r \neq 0. \qquad (1)$$

Here r can be any integer, positive, negative or zero. From the definition of r as $v(c)$ it is clear that $r \geqslant 0$ precisely when $c \in V$.

We remark that the field structure of K is entirely determined by the expression (1) for its elements, together with addition and multiplication tables for the transversal A. By this we mean the following: given $a, a' \in A$, the elements $a + a'$ and aa' lie in V and so are congruent (mod π) to uniquely determined elements of A:

$$a + a' \equiv s, \qquad aa' \equiv p \pmod{\pi}, \text{ where } s, p \in A. \tag{2}$$

These congruences for all $a, a' \in A$ just constitute the addition and multiplication tables for A, and provide a presentation for V. With their help we can add and multiply any expressions (1). Taking $r = 0$ for convenience, suppose that we have $c = \Sigma a_i\pi^i$ and $c' = \Sigma a'_i\pi^i$; then $c + c' = \Sigma s_i\pi^i$, where $s_0, s_1, \ldots$ are the unique elements of A determined by

$$s_0 \equiv a_0 + a'_0 \pmod{\pi},$$

$$s_1 \equiv (a_0 + a'_0 - s_0)\pi^{-1} + a_1 + a'_1 \pmod{\pi},$$

$$\ldots .$$

Similarly, $cc' = \Sigma p_i\pi^i$, where $p_0, p_1, \ldots$ are the unique elements of A determined by

$$p_0 \equiv a_0 a'_0 \pmod{\pi},$$

$$p_1 \equiv (a_0 a'_0 - p_0)\pi^{-1} + a_0 a'_1 + a_1 a'_0 \pmod{\pi},$$

$$\ldots ,$$

analogous to the formulae for 'carrying' digits in the addition and multiplication of decimal fractions.

Since (1) holds for all $n \geqslant r$, we may represent c by the infinite series

$$c = \sum_r^\infty a_i\pi^i, \tag{3}$$

where this is understood to mean the conjunction of all the congruences (1) for $n = r, r + 1, \ldots$. We note that the infinite series determines c uniquely, for if $\Sigma a_i\pi^i$ also represents c', then by taking the series as far as π^n we see that $v(c - c') \geqslant n$, and this holds for all n, therefore $v(c - c') = \infty$ and $c' = c$. Thus every element of K is represented by an infinite series (3), which in turn determines the element completely.

However, it is not generally true that every series (3) represents an element of K; in fact this holds precisely when K is complete. For any sequence of elements of the form

$$c_n = \sum_r^n a_i\pi^i \quad (n = r, r + 1, \ldots) \tag{4}$$

is a Cauchy sequence, because $v(c_m - c_n) \geqslant \min\{m, n\}$; so if K is complete, then equation (4) converges to $\Sigma_r^\infty a_i\pi^i$ and therefore every series (3) represents an element of K. Conversely, if every series (3) represents an element of K and we have a Cauchy sequence (s_n), we can expand every element s_n in the form (3); since $v(s_m - s_n) \to \infty$, the coefficient of any fixed power π^h must be the same for all s_n for $n > n_0(h)$. If the ultimate value of this coefficient is a_h, then the sequence s_n converges to $\Sigma a_h\pi^h$. The details are straightforward and may be left to the reader. We sum up our result as follows:

Theorem 5.1

Let K be a field with a principal valuation v, assumed normalized, denote by π a uniformizer and by A a transversal for the residue class field of K, to include 0. Then every element c of $K^\times$ can be uniquely represented in the form

$$c = \Sigma_r^\infty a_i \pi^i \quad \text{where} \quad a_i \in A,\ r = v(c),\ a_r \neq 0. \tag{5}$$

The addition and multiplication in K are defined by the formulae (2) and further, every expression (5) represents an element of K if and only if K is complete. ■

This result makes it of interest to choose the transversal A so as to make the addition and multiplication as simple as possible. For a complete field of characteristic equal to that of its residue class field and where the latter is perfect, we can choose A to form a field (cf. the second example below); this makes the formulae particularly simple. In the case of unequal characteristic this is clearly impossible, but one can still choose A to be multiplicative. As we have no occasion to use this representation we do not give the details here (cf. Hasse, (1980), Chapter 10, and Exercise 5 below).

As an example, let us apply Theorem 5.1 to $\mathbf{Q}_p$, the p-adic completion of the rational numbers. As transversal for the residue class field we take $A = \{0, 1, \ldots, p-1\}$. We see that every non-zero rational number can be written as a p-adic series

$$c = \sum_r^\infty a_i p^i, \quad a_i = 0, 1, \ldots, p-1; \quad a_r \neq 0, \tag{6}$$

and the set of all such series, together with 0, constitutes the completion $\mathbf{Q}_p$. The ring $\mathbf{Z}_p$ of p-adic integers in $\mathbf{Q}_p$ is the set of all series (6) with $r \geqslant 0$. The subfield of rational numbers can be characterized as the set of all series (6) that are ultimately periodic (as in the case of decimal fractions):

Proposition 5.2

The element c of $\mathbf{Q}_p$ represented by (6) is rational if and only if, for some r_0, t and all $i \geqslant r_0$, $a_{i+t} = a_i$, i.e. the sequence (a_i) is ultimately periodic.

Proof

If the series (6) is ultimately periodic, then c can be written as the sum of a finite series and one with periodic coefficients. The finite part is clearly rational, while the periodic part has the form

$$\left(\sum_0^{t-1} a_i p^i\right)(1 + p^t + p^{2t} + \ldots) = \sum_0^{t-1} a_i p^i (1 - p^t)^{-1}.$$

This proves the sufficiency of the condition.

Conversely, given any rational number $c = p^r \cdot m/n$, where $p \nmid mn$, we note that p is a unit in the multiplicative group of $\mathbf{Z}/n$, hence $p^t \equiv 1 \pmod{n}$ for some t. Suppose that $p^t - 1 = nn'$; then

$$p^r m/n = -p^r mn'(1-p^t)^{-1}.$$

Here we can divide mn' by $p^t - 1$ with a negative remainder:

$$mn' = a(p^t - 1) - b, \qquad \text{where } 0 < b < p^t - 1.$$

$$\text{Now } c = p^r m/n = p^r mn'(p^t - 1)^{-1}$$

$$= p^r a + p^r b(1-p^t)^{-1},$$

and on expansion this yields a series which is ultimately periodic.■

As a second example we take the rational function field $k(x)$. As we have seen in Proposition 4.4, every non-trivial valuation v of $k(x)$ over k is principal. Suppose first that $v(x) \geqslant 0$; then the uniformizer is an irreducible polynomial f, of degree n say, over k, and the residue class field has the form $k(\alpha)$, where α is a zero of f. As transversal we can take the set A of all polynomials in x over k of degree less than n. Thus each element ϕ of $k(x)$ has the form

$$\phi = \sum_r^\infty a_i f^i, \quad a_i \in A,\ v(\phi) = r,\ a_r \neq 0. \tag{7}$$

Of course this also follows easily by a direct argument: we can expand ϕ as a formal Laurent series in x and then reduce the nth and higher powers of $x \bmod f$. Now it is clear that the completion of $k(x)$ consists of all formal series (7) and 0. The field of formal Laurent series in x over k is written $k((x))$.

Consider for a moment the special case where $n = 1$. Then we may take f in the form $f = x - \alpha$; the residue class field is just k and the completion consists of all series

$$\phi = \sum_r^\infty a_i (x-\alpha)^i, \quad a_i \in k,\ v(\phi) = r,\ a_r \neq 0. \tag{8}$$

If $r \geqslant 0$, then we can put $x = \alpha$ and obtain the value a_0 for ϕ; this is the value of ϕ at the place $x = \alpha$. We cannot put x equal to any other value; if $x = \beta \neq \alpha$, we get $\Sigma a_i(\beta - \alpha)^i$ which has no meaning in general, because we have no notion of convergence in k. Going back to the case where $n > 1$, we can of course (for $r \geqslant 0$) put $x = \alpha$ in (7), but now α is not in k. We thus obtain the value a_0 for ϕ at the place $x = \alpha$, which is algebraic over k.

There remains the case when $v(x) < 0$. Now x^{-1} is a uniformizer and we obtain an expansion as an inverse power series

$$\phi = \sum_r^\infty b_i x^{-i}, \quad b_i \in k,\ v(\phi) = r,\ b_r \neq 0.$$

When $r \geqslant 0$, we obtain b_0 as the value of ϕ at the place $x = \infty$.

Exercises

1. Which rational numbers can be written p-adically in purely periodic form?
2. The representation as decimal fraction is not unique, since e.g. 1.000 . . . = 0.999 . . . . Why is this ambiguity absent in the p-adic case?
3. Show that (8) is a rational function of x if and only if there exist integers r, n_0 and $c_1, \ldots, c_r \in k$ such that

$$a_n = a_{n-1}c_1 + a_{n-2}c_2 + \ldots + a_{n-r}c_r \qquad \text{for all } n > n_0.$$

4. Show that every $(p-1)$th root of 1 in $\mathbf{Z}_p$ is of the form $\lim_{r\to\infty}\left(a^{p^r}\right)$, where $0 < a < p$.
5. Show that for any odd prime p, every root of 1 in $\mathbf{Q}_p$ is a $(p-1)$th root of 1, while in $\mathbf{Q}_2$ every root of 1 is a square root of 1. Show that the roots of 1 form a multiplicative transversal in $\mathbf{Q}_p$.
6. Let $I = [0,1]$ be the unit interval with the usual topology (induced from $\mathbf{R}$). Show that every continuous map from I to $\mathbf{Z}_p$ is constant; show also that the map

$$\sum a_i p^i \mapsto \sum a_i p^{-i-1}$$

from $\mathbf{Z}_p$ to I is continuous. Is it injective or surjective?

1.6 ORDERED GROUPS

In this section and the next we take a brief look at valuations of higher rank. These results are mainly used in Chapter 5, when we come to classify the valuations on a function field in two variables, so the reader may wish merely to skim these sections and return to them later, but apart from their importance in the general theory these notions also help to illustrate the special case with which we are concerned here.

Let Γ be an ordered abelian group, i.e. an abelian group which has a total ordering which is preserved by the group operation. We shall usually write Γ additively and call an element α of Γ *positive* if $\alpha > 0$. A subset Σ of Γ is said to be *convex* if $\alpha, \beta \in \Sigma$, say $\alpha \leqslant \beta$, and $\alpha \leqslant \lambda \leqslant \beta$ for some $\lambda \in \Gamma$ implies $\lambda \in \Sigma$. In particular, we shall be concerned with convex subgroups (sometimes called 'isolated' subgroups). We note that the whole group and the trivial subgroup $\{0\}$ are always convex. If Δ is a convex subgroup of Γ, then each coset of Δ is convex and the cosets define a partition of Γ respecting the order; this allows the quotient group Γ/Δ to be defined as an ordered group in a natural way which is made precise in Theorem 6.1 below. By an *order-homomorphism* of ordered groups we understand a homomorphism respecting the order; *order-endomorphism, order-embedding* and *order-automorphism* are defined similarly.

Theorem 6.1
For any order-homomorphism $f: \Gamma \to \Gamma'$ of ordered abelian groups, the kernel $\ker f$ is a convex subgroup of Γ. Conversely, given a convex subgroup Δ of Γ, the quotient group Γ/Δ has a unique ordered group structure such that the natural mapping $\Gamma \to \Gamma/\Delta$ is an order-homomorphism.

Proof
Let $\alpha, \beta \in \ker f$, say $\alpha \leqslant \beta$ and take $\lambda \in \Gamma$ such that $\alpha \leqslant \lambda \leqslant \beta$. Then

$$0 = \alpha f \leqslant \lambda f \leqslant \beta f = 0,$$

hence $\lambda f = 0$ and so $\lambda \in \ker f$; this shows $\ker f$ to be convex. Conversely, given a convex subgroup Δ, we easily see that each coset $\Delta + \alpha$ is convex; if $\alpha \mapsto \bar{\alpha}$ is the natural homomorphism and we put $\bar{\alpha} \leqslant \bar{\beta}$ if and only if $\alpha \leqslant \beta$, then this defines an ordered group structure on Γ/Δ, and it is clearly the only ordering for which the natural map is an order-homomorphism.■

The next result is a fairly obvious remark which is nevertheless quite basic. For any α in an ordered group we define

$$|\alpha| = \begin{cases} \alpha & \text{if } \alpha \geqslant 0, \\ -\alpha & \text{if } \alpha < 0. \end{cases}$$

Thus $|\alpha| \geqslant 0$ for all α in the group. Further, we shall say that α *majorizes* β if $|\beta| \leqslant |\alpha|$. It is clear that for any α, β, one majorizes the other, and a convex subgroup contains all elements majorized by one of its members.

Theorem 6.2
In any ordered group the set of convex subgroups is totally ordered by inclusion.

Proof
Let Δ_1, Δ_2 be convex subgroups of an ordered group Γ and assume that $\Delta_1 \not\subseteq \Delta_2$, say $\alpha \in \Delta_1$, $\alpha \notin \Lambda_2$. Then no element of Δ_2 can majorize α, for otherwise α would lie in Δ_2, by convexity. Hence α majorizes each element of Δ_2 and so $\Delta_2 \subseteq \Delta_1$.■

The order-type of the chain of proper convex subgroups of Γ is called the *rank* of Γ, written rk Γ. We shall mostly be concerned with groups of rank 1; they can also be characterized in other ways which we list next. In any totally ordered set S, a subset L is called a *lower segment* if $\alpha \in L$ and $\beta \leqslant \alpha$ implies $\beta \in L$. *Upper segments* are defined by symmetry, or as complements of lower segments. A familiar method of constructing **R** defines each real number as a pair consisting of a lower segment and its complementary upper segment in **Q** (a 'Dedekind cut', where a convention is made to take the rational division point, if any, to the lower segment); this is Dedekind's method of completion by cuts, to construct the real numbers.

Proposition 6.3
Let Γ be an abelian ordered group. Then Γ has positive rank if and only if it is non-trivial; further, the following conditions are equivalent:

(*a*) *Γ has rank at most 1;*
(*b*) *Γ is archimedean ordered, i.e. given $\alpha, \beta \in \Gamma$, $\alpha > 0$, there exists $n \in \mathbf{N}$ such that $n\alpha > \beta$;*
(*c*) *Γ has an order-embedding in $\mathbf{R}^+$, the additive group of real numbers.*

Proof
It is clear that Γ has rank 0 if and only if $\Gamma = \{0\}$, so we may take the rank to be positive.

(a) $\Rightarrow$ (b). Assume that rk $\Gamma = 1$. Given $\alpha, \beta \in \Gamma$, if $\alpha > 0$, then the convex subgroup generated by α is the whole of Γ, for it is the set of all elements λ such that $-n\alpha \leqslant \lambda \leqslant n\alpha$ for some n. Thus $\beta \leqslant n\alpha$ for some n and (b) holds.

(b) $\Rightarrow$ (c). Fix $\lambda > 0$ in Γ and for any $\alpha \in \Gamma$ define

$$L(\alpha) = \left\{ \frac{m}{n} \,\middle|\, m\lambda \leqslant n\alpha \right\}, \qquad U(\alpha) = Q \backslash L(\alpha).$$

By (b), $L(\alpha), U(\alpha) \neq \varnothing$. Moreover, if $m/n \in L(\alpha)$ and $m'/n' \leqslant m/n$, then taking $n, n' > 0$, say, we have $m'n \leqslant mn'$. Since $m\lambda \leqslant n\alpha$, we have $mn'\lambda \leqslant nn'\alpha$, hence

$$m'n\lambda \leqslant mn'\lambda \leqslant nn'\alpha$$

and therefore

$$m'/n' = m'n/n'n \in L(\alpha).$$

This shows $L(\alpha)$ to be a lower segment in $\mathbf{Q}$; it follows that its complement $U(\alpha)$ is an upper segment. We thus have a Dedekind cut. Let us define $\phi(\alpha)$ as the corresponding real number, thus $L(\alpha)$ consists of the rational numbers $\leqslant \phi(\alpha)$ and $U(\alpha)$ of the rational numbers $> \phi(\alpha)$. We claim that $\alpha \mapsto \phi(\alpha)$ is an order-embedding of Γ in $\mathbf{R}$.

Given $\alpha, \alpha' \in \Gamma$, if $m/n \in L(\alpha), m'/n' \in L(\alpha')$, then

$$m\lambda \leqslant n\alpha, \; m'\lambda \leqslant n'\alpha',$$

hence

$$mn'\lambda \leqslant nn'\alpha, \; m'n\lambda \leqslant nn'\alpha'$$

and so

$$(mn' + m'n)\,\lambda \leqslant nn'(\alpha + \alpha'),$$

therefore

$$m/n + m'/n' \in L(\alpha + \alpha').$$

Thus for any $r, r' \in \mathbf{Q}$, $r \leqslant \phi(\alpha)$, $r' \leqslant \phi(\alpha') \Rightarrow r + r' \leqslant \phi(\alpha + \alpha')$. Letting $r \to \phi(\alpha)$, $r' \to \phi(\alpha')$ we find that

$$\phi(\alpha) + \phi(\alpha') \leqslant \phi(\alpha + \alpha').$$

Similarly we have

$$\phi(\alpha + \alpha') \leqslant \phi(\alpha) + \phi(\alpha');$$

thus

$$\phi(\alpha + \alpha') = \phi(\alpha) + \phi(\alpha')$$

and this shows ϕ to be a homomorphism. Clearly it is order-preserving and $\phi(\lambda) = 1$. If $\alpha \neq 0$, say $\alpha > 0$, then $n\alpha > \lambda$ for some n, hence $\phi(\alpha) > 1/n$ and so $\phi(\alpha) \neq 0$, hence ϕ is an order-embedding of Γ in $\mathbf{R}$.

Now (c) $\Rightarrow$ (a) is clear: the convex subgroup of $\mathbf{R}$ generated by any positive number is $\mathbf{R}$ itself, so $\mathbf{R}$ has rank 1 and the same still holds for any non-trivial subgroup.■

We observe that the mapping ϕ constructed in Proposition 6.3 is completely determined once λ has been chosen. Now let f be any order-endomorphism of $\mathbf{R}$ and put $f(1) = \lambda$; then λ determines f completely. But the map $\alpha \mapsto \alpha\lambda$ is an order-endomorphism of $\mathbf{R}$ mapping α to $\alpha\lambda$ and so must coincide with f:

Corollary 6.4

Any order-endomorphism of $\mathbf{R}^+$ has the form $f: \alpha \mapsto \alpha\lambda$, for some fixed $\lambda \geqslant 0$. In particular, f is either an automorphism or 0.

The last part is clear from the explicit form of f; of course it also follows from Theorem 6.1 and the fact that rk $\mathbf{R} = 1$.■

An ordered group is said to be *discrete* if the set of all convex subgroups is well-ordered and in each non-trivial order-homomorphic image each element has an immediate successor. Let us consider a discrete group Γ_0 of rank 1. If λ is the least positive element, then for any $\alpha \in \Gamma_0$ there exists $n \in \mathbf{N}$ such that $(n+1)\lambda > \alpha$. If $\alpha > 0$ and we choose the least such n, then $n\lambda \leqslant \alpha$, so $0 \leqslant \alpha - n\lambda < \lambda$. By the definition of λ it follows that $\alpha - n\lambda = 0$, so $\alpha = n\lambda$. Similarly, if $\alpha < 0$, then $\alpha = -n\lambda$ for some $n \in \mathbf{N}$, hence $\Gamma_0 \cong \mathbf{Z}$. Thus we have established:

Proposition 6.5

A discrete ordered group of rank 1 is isomorphic to $\mathbf{Z}$, and $\mathbf{Z}$ has no order-automorphism other than the identity.

The last part follows because any order-automorphism must map the positive generator to itself.■

As an example of a discrete ordered group of rank n we have $\mathbf{Z}^n$, the group of n-tuples of integers, with componentwise addition and with the lexicographic ordering:

$$(\alpha_1, \ldots, \alpha_n) \leqslant (\beta_1, \ldots, \beta_n)$$

holds precisely if the first non-zero difference $\beta_i - \alpha_i$ is positive. We shall now show that every discrete ordered abelian group of finite rank has this form:

Theorem 6.6
Any discrete ordered abelian group Γ of finite rank n is of the form $\mathbf{Z}^n$, with the lexicographic ordering.

Proof
Consider the chain of convex subgroups in Γ:

$$\Gamma = \Gamma_0 \supset \Gamma_1 \supset \ldots \supset \Gamma_n = 0. \tag{1}$$

Each quotient Γ_{i-1}/Γ_i is discrete of rank 1, hence isomorphic to $\mathbf{Z}$. In the coset which is the canonical generator of Γ_{i-1}/Γ_i pick an element e_i, so that $\Gamma_{i-1} = \Gamma_i + \langle e_i \rangle$. This sum is direct, because Γ_{i-1}/Γ_i is free on the coset $\Gamma_i + e_i$. Hence by induction on n, Γ is the free abelian group on $e_1, \ldots, e_n$ as basis. Moreover, if

$$\sum \alpha_i e_i > \sum \beta_i e_i, \tag{2}$$

then by passing to Γ_0/Γ_1 we see that $\alpha_1 \geqslant \beta_1$. If $\alpha_1 = \beta_1$, we subtract $\alpha_1 e_1$ on both sides of (2) and obtain an expression in Γ_1. Now $\alpha_2 \geqslant \beta_2$ and continuing in this way we see that (2) holds precisely when the first non-zero difference $\alpha_i - \beta_i$ is positive.■

For $n = 1$ we have seen that $\mathbf{Z}$ has no order-automorphism $\neq 1$, but there are such automorphisms when $n > 1$. Taking $n = 2$, let f be an order-automorphism of $\mathbf{Z}^2$. The least positive element, (0,1), maps to itself, while (1, 0) maps to $(1, c)$. This shows that

$$f(x_1, x_2) = x_1(1, c) + x_2(0, 1) = (x_1, x_1 c + x_2),$$

and such a mapping, represented by the matrix

$$\begin{pmatrix} 1 & c \\ 0 & 1 \end{pmatrix},$$

is clearly an order-automorphism for any $c \in \mathbf{Z}$. Hence the group of all order-automorphisms of $\mathbf{Z}^2$ is isomorphic to $\mathbf{Z}$.

For a later application (in Chapter 5) we briefly consider non-discrete subgroups of $\mathbf{Q}$. To classify them we need a definition. By a *supernatural number* or *Steinitz number* we understand a formal product

$$M = \prod p_i^{\alpha_i}, \quad p_i \text{ primes}, \ \alpha_i = 0, 1, 2, \ldots \text{ or } \infty.$$

Thus M is a product of any set of prime powers, allowing as exponents positive integers or ∞. If $N = \prod p_i^{\beta_i}$ is a second such number, we write $M|N$ to mean: $\alpha_i \leqslant \beta_i$ for all i (where it is understood that $n \leqslant \infty$ for $n = 0, 1, 2, \ldots$). It is clear that these numbers include the natural numbers as a special case: all the α_i are finite and almost all are 0. We can define the highest common factor (HCF) and least common multiple (LCM) of M and N as

$$\mathrm{HCF}\,(M, N) = \prod p_i^{\min(\alpha_i,\, \beta_i)}, \qquad \mathrm{LCM}\,(M, N) = \prod p_i^{\max(\alpha_i,\, \beta_i)}.$$

We remark that any supernatural number is completely determined by its natural divisors. In fact supernatural numbers are simply a convenient way of describing sets S of natural numbers with the properties:

1. $a \in S, b \mid a \Rightarrow b \in S$;
2. $a, b \in S \Rightarrow LCM(a, b) \in S$.

Their relevance for us is that they are precisely the sets occurring as denominator sets of subgroups of $\mathbf{Q}^+$:

Proposition 6.7
Any non-trivial subgroup Γ *of* $\mathbf{Q}^+$ *has the form*

$$\Gamma_N = \{r/s \mid s \mid N\} \quad \textit{for some supernatural number } N.$$

Moreover, $\Gamma_M \cong \Gamma_N$ *if and only if* $N = \alpha M$ *for some* $\alpha \in \mathbf{Q}^\times$.

Proof
By hypothesis, $\Gamma \neq 0$, so by a suitable normalization we may assume that $1 \in \Gamma$. If Γ contains r/s, where r, s are coprime, then Γ contains $1/s$; for there exist $a, b \in \mathbf{Z}$ such that $ar + bs = 1$, hence $s^{-1} = ar/s + b \in \Gamma$. Next, if Γ contains $1/s'$ and $1/s''$ then Γ contains $1/m$, where m is the LCM of s' and s''. For let $m = s't' = s''t''$; then t', t'' are coprime, so $ut' + vt'' = 1$ for some $u, v \in \mathbf{Z}$ and hence

$$m^{-1} = (ut' + vt'')/m = u/s' + v/s'' \in \Gamma.$$

It follows that the denominators make up a supernatural number N and $\Gamma \cong \Gamma_N$.

This isomorphism is uniquely determined, once we have chosen an element of Γ to correspond to $1 \in \Gamma_N$. If $\Gamma_M \cong \Gamma_N$, let $\alpha \in \Gamma_M$ correspond to $1 \in \Gamma_N$; then for any $n \mid N$, $1/n \in \Gamma_N$, hence $\alpha/n \in \Gamma_M$ and all denominators in Γ_N occur in this way, hence $M = \alpha^{-1}N$, i.e. $N = \alpha M$. ∎

To describe a general non-discrete subgroup of $\mathbf{R}$ one picks a maximal family of elements (e_i) linearly independent over $\mathbf{Q}$; the numbers occurring as coefficient of a given e_i form a subgroup of $\mathbf{Q}$, to which Proposition 6.7 can be applied. We omit the details, since they will not be needed.

For any ordered group Γ and a convex subgroup Δ we have the formula

$$rk\,\Gamma = rk\,\Delta + rk\,\Gamma/\Delta, \tag{3}$$

which is easily established, using the chain (1). Sometimes we shall need to find the rank of a non-convex subgroup; the following is the only case needed here:

Proposition 6.8
If Γ is any ordered group and Γ' any subgroup such that Γ/Γ' is a torsion group, then Γ and Γ' have the same rank.

Proof
We claim that the correspondence

$$\Delta \mapsto \Delta \cap \Gamma' \tag{4}$$

provides a bijection between the set of convex subgroups of Γ and those of Γ'.

It is clear that $\Delta \cap \Gamma'$ is convex in Γ' whenever Δ is convex in Γ. If Δ' is any convex subgroup of Γ' and Δ is the convex subgroup of Γ generated by Δ', then Δ consists of all elements majorized by an element of Δ' and so $\Delta \cap \Gamma' = \Delta'$, because Δ' was convex. This shows (4) to be surjective. If

$$\Delta_1 \cap \Gamma' = \Delta_2 \cap \Gamma'$$

and $\alpha \in \Delta_1$, then for some $m > 0$,

$$m\alpha \in \Delta_1 \cap \Gamma' = \Delta_2 \cap \Gamma',$$

but $m\alpha$ majorizes α, so $\alpha \in \Delta_2$. This shows that $\Delta_1 \subseteq \Delta_2$, and by symmetry we have $\Delta_1 = \Delta_2$. Thus (4) is indeed a bijection, and it follows that $rk\,\Gamma' = rk\,\Gamma$. ■

Exercises

1. Show that an abelian group can be defined as an ordered group precisely when it is torsion free.
2. Show that a non-abelian totally ordered group which is archimedean ordered (in the sense of Proposition 6.3 (b)) has an order-embedding in **R** (and so is abelian after all).
3. Describe the order-automorphisms of $\mathbf{Z}^n$.
4. Determine all discrete ordered groups (not necessarily abelian) of rank 2.
5. Prove an analogue of Theorems 6.1 and 6.2 for non-abelian groups.
6. Describe the non-discrete subgroups of **R** with exactly two Q-linear independent elements (such a group is sometimes said to have *rational rank* 2).

1.7 GENERAL VALUATIONS

Let K be a field with a general valuation v, with valuation ring V, maximal ideal $\mathfrak{m}$ and residue class field $K_v = V/\mathfrak{m}$. A second valuation v' on K, with valuation ring V', is said to be *subordinate* to v, $v' \preccurlyeq v$, if $V' \supseteq V$. If the maximal ideal of V' is $\mathfrak{m}'$ and the unit groups of V, V' are U, U' respectively, then it is clear that $U' \supseteq U$, $\mathfrak{m}' \subseteq \mathfrak{m}$, and we can illustrate the situation by the accompanying diagram. In the natural homomorphism $V' \to V'/\mathfrak{m}' = K_{v'}$ the valuation ring V corresponds to a valuation ring in $K_{v'}$ with residue class field K_v. Thus the place $K \to K_{v'}$ can be composed with the place $K_{v'} \to K_v$ to give the place $K \to K_v$. Conversely, any valuation on $K_{v'}$, can be composed with v', giving a valuation on K to which v' is subordinate.

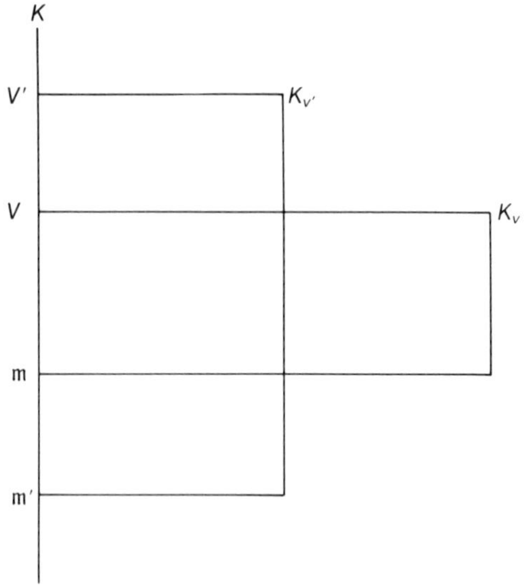

Further, if the valuations v, v' have the value groups Γ, Γ' respectively, then

$$\Gamma = K^{\times}/U,$$
$$\Gamma' = K^{\times}/U' \cong (K^{\times}/U) \,/\, (U'/U) = \Gamma/\Delta,$$

where $\Delta \cong U'/U$ is a convex subgroup of Γ. Thus if $\phi : \Gamma \to \Gamma'$ is the natural homomorphism, then $v'(x) = \phi(v(x))$ for all $x \in K$. In this way we can decompose the valuation v' as ϕv and we see that $K_{v'}$ has a valuation with value group $\Delta = \ker \phi$ and residue class field K_v.

We recall that for any prime ideal $\mathfrak{p}$ in a commutative ring R the complement $R\backslash\mathfrak{p}$ is multiplicatively closed and we can define the *localization* of R at $\mathfrak{p}$ as the ring of fractions

$$R_{\mathfrak{p}} = \{a/u \mid a, u \in R, \quad u \notin \mathfrak{p}\}.$$

This is easily seen to be a *local ring*, i.e. the non-units form the unique maximal ideal, and this maximal ideal corresponds to $\mathfrak{p}$ in the natural homomorphism $R \to R_{\mathfrak{p}}$ (cf. A.2, section 9.3). We remark that the passage from R to $R_{\mathfrak{p}}$ is accomplished by 'making all elements outside $\mathfrak{p}$ invertible'; this means that the ideals contained in $\mathfrak{p}$ are unaffected while the others disappear under localization. It follows that there is an order-preserving bijection between the ideals of $R_{\mathfrak{p}}$ and the ideals of R contained in $\mathfrak{p}$.

We now show how the localization process can be applied to form the valuation ring of a subordinate valuation. Let a field K be given, with a valuation ring V and its maximal ideal $\mathfrak{m}$. It is clear that each ring V' such that $V \subseteq V' \subset K$ is a valuation ring in K, with maximal ideal $\mathfrak{m}'$ satisfying $0 \subset \mathfrak{m}' \subseteq \mathfrak{m}$; moreover, V' can be obtained from V by localizing at $\mathfrak{m}'$:

$$V' \cong V_{\mathfrak{m}'}.$$

Conversely, given any prime ideal $\mathfrak{p}$ in V, the localization $V_{\mathfrak{p}}$ is a valuation ring in K containing V, in which the maximal ideal corresponds to $\mathfrak{p}$. Thus the valuations subordinate to v correspond to the rings containing V (necessarily valuation rings), and to the prime ideals of V. We can also define an inclusion-reversing bijection between the set of all prime ideals of V and the set of convex subgroups of Γ, the value group of v. For any prime ideal $\mathfrak{p}$ in V we put

$$\mathfrak{p}^* = \{\lambda \in \Gamma \mid |\lambda| < v(x) \text{ for all } x \in \mathfrak{p}\}.$$

Clearly $\mathfrak{p}^*$ is a convex subset of Γ; it is a subgroup, for if $\alpha, \beta \in \mathfrak{p}^*$, take $x, y \in V$ such that $v(x) = |\alpha|$, $v(y) = |\beta|$. Then $x, y \notin \mathfrak{p}$, so $xy \notin \mathfrak{p}$, because $\mathfrak{p}$ is prime, and hence

$$|\alpha + \beta| \leqslant |\alpha| + |\beta| = v(x) + v(y) = v(xy).$$

Any z in $\mathfrak{p}$ satisfies $v(z) > v(xy) \geqslant |\alpha + \beta|$, therefore $\alpha + \beta \in \mathfrak{p}^*$. It is clear that $0 \in \mathfrak{p}^*$ and $\alpha \in \mathfrak{p}^*$ implies $-\alpha \in \mathfrak{p}^*$, so $\mathfrak{p}^*$ is indeed a convex subgroup of Γ.

Conversely, given any convex subgroup Δ of Γ, we define

$$\Delta^* = \{x \in V \mid v(x) > |\lambda| \text{ for all } \lambda \in \Delta\}.$$

Then Δ^* is an ideal in V, for it is of the form $\cap \mathfrak{a}_\lambda$, where

$$\mathfrak{a}_\lambda = \{x \in V \mid v(x) > |\lambda|\}.$$

It is prime since $1 \notin \Delta^*$ and if $x, y \notin \Delta^*$, then $v(x), v(y) \in \Delta$, hence

$$v(xy) = v(x) + v(y) \in \Delta$$

and so $xy \notin \Delta^*$.

It is clear that $\mathfrak{p}_1 \subseteq \mathfrak{p}_2 \Rightarrow \mathfrak{p}_1^* \supseteq \mathfrak{p}_2^*$ and $\Delta_1 \subseteq \Delta_2 \Rightarrow \Delta_1^* \supseteq \Delta_2^*$. To show that we have a bijection we note that $\mathfrak{p}^{**} \supseteq \mathfrak{p}$, $\Delta^{**} \supseteq \Delta$. Now if $x \notin \mathfrak{p}$, then $v(x) \in \mathfrak{p}^*$ and so $x \notin \mathfrak{p}^{**}$, therefore $\mathfrak{p}^{**} = \mathfrak{p}$. Similarly, if $\lambda \notin \Delta$, then λ majorizes all the elements of Δ. Take $x \in V$ to satisfy $v(x) = |\lambda|$; then $|\mu| < v(x)$ for all $\mu \in \Delta$, hence $x \in \Delta^*$ and so $v(x) \notin \Delta^{**}$, therefore also $\lambda \notin \Delta^{**}$, and so $\Delta^{**} = \Delta$. We sum up our findings as follows:

Theorem 7.1

Let K be a field with a general valuation v, with valuation ring V and value group Γ. Then there is an inclusion-reversing bijection between (i) valuation rings containing V and prime ideals in V, and (ii) prime ideals in V and convex subgroups of Γ.

Hence there is an inclusion-preserving bijection between valuation rings containing V and convex subgroups of Γ. Further, if V', Δ are a valuation ring and convex subgroup corresponding under these bijections, then the value group for the valuation corresponding to V' is Γ/Δ.■

Since the set of all convex subgroups is totally ordered by inclusion, by Proposition 6.2, we have the following consequence:

Corollary 7.2

The set of valuation rings above a given valuation ring is totally ordered by inclusion.■

Theorem 7.1 also leads to a description of rank 1 valuations, in terms of the valuation ring:

Corollary 7.3

A valuation v on a field K has rank 1 if and only if its valuation ring is maximal among subrings of K different from K.

For if V denotes the valuation ring and Γ the value group of v, then by Theorem 7.1 the convex subgroups of Γ correspond to the valuation rings above V, so V is maximal precisely when Γ has no non-trivial convex subgroups, i.e. it is of rank 1.■

The above analysis also allows us to give another characterization of rank 1 valuation rings, which will be useful later on:

Proposition 7.4

Let K be a field and A a subring of K. Then the following conditions are equivalent:

(a) *A is the valuation ring of a rank 1 valuation on K;*
(b) *A is a valuation ring in K with a single non-zero prime ideal;*
(c) *A is not a field and is a maximal proper subring of K.*

Proof

(a) ⇒ (b). Assume that (b) is false, thus A has a non-zero prime ideal $\mathfrak{p}$ other than the maximal ideal. Then the localization $A_\mathfrak{p}$ properly contains A: $A_\mathfrak{p} \supset A$; if the valuations corresponding to A, $A_\mathfrak{p}$ have value groups Γ, Γ' respectively, then $\Gamma' \cong \Gamma/\Delta$, where Δ is a convex subgroup of Γ, the value group of the valuation on the residue class field for v (cf. the diagram on p.38). Hence $\operatorname{rk}\Gamma = \operatorname{rk}\Gamma' + \operatorname{rk}\Delta \geqslant 2$ and so (a) is false.

(b) $\Rightarrow$ (c). Let A be as in (b) and suppose that B is a subring of K such that $A \subset B \subset K$. Clearly B is again a valuation ring; let $\mathfrak{m}$, $\mathfrak{n}$ be the maximal ideals of A, B respectively. Then $\mathfrak{p} = A \cap \mathfrak{n}$ is a non-zero prime ideal in A and if $u \in B \backslash A$, then $u^{-1} \in \mathfrak{m} \backslash \mathfrak{n}$ hence $\mathfrak{p} \subset \mathfrak{m}$. This contradicts (b), so A must be maximal.

(c) $\Rightarrow$ (a). Assume that A is as in (c) and let $c \in A$, $c^{-1} \notin A$. Then $A[c^{-1}]$ is a ring strictly containing A, hence $A[c^{-1}] = K$. Now K cannot be finitely generated as an A-module, for if it were, then every element of K would have the form ac^{-m} for $a \in A$ and a fixed m. Thus $c^{-m-1} = ac^{-m}$ and hence $c^{-1} = a \in A$, a contradiction.

To show that A is a valuation ring, suppose that $u, u^{-1} \notin A$. Then $A[u] = A[u^{-1}] = K$, so

$$u = a_0 + a_1 u^{-1} + \ldots + a_r u^{-r},$$

and it follows that $K = A[u]$ is finitely generated as A-module, by 1, $u^{-1}, \ldots, u^{-r}$, which contradicts what was proved earlier. Hence u or u^{-1} is in A, which is therefore a valuation ring. To show that it has rank 1, let $c \in A$, $c^{-1} \notin A$. Then any $x \in K$ has the form $x = ac^{-m}$ for $a \in A$ and some m, so the valuation satisfies $v(x) \geqslant -mv(c)$, and this shows it to have rank 1, as claimed.■

We conclude this section with an example to show that valuated fields with an arbitrary value group exist. Let Γ be any ordered abelian group, k any field and consider the *group algebra* $k\Gamma$, i.e. the k-space on Γ as basis, with multiplication defined by the group operation in Γ. Since Γ was written additively, we shall use exponents and write the general element of $k\Gamma$ as a finite sum

$$f = \sum_{\alpha \in \Gamma} a_\alpha x^\alpha, \tag{1}$$

where $a_\alpha \in k$ and almost all the a_α vanish. The multiplication is given by the rule

$$x^\alpha x^\beta = x^{\alpha+\beta}. \tag{2}$$

We define v on $k\Gamma$ by putting $v(0) = \infty$ and for $f \neq 0$, given by (1) say, put

$$v(f) = \min\{\alpha \in \Gamma \mid a_\alpha \neq 0\}.$$

This indeed defines a valuation on $k\Gamma$, for if $f = a_\alpha x^\alpha + \ldots$, $g = b_\beta x^\beta + \ldots$, where a_α, $b_\beta \neq 0$ and dots indicate higher terms, then

$$fg = a_\alpha b_\beta x^{\alpha+\beta} + \ldots,$$

so

$$v(fg) = \alpha + \beta = v(f) + v(g),$$

and

$$f + g = a_\alpha x^\alpha + \ldots + b_\beta x^\beta + \ldots,$$

so $v(f+g) = \min\{\alpha, \beta\}$ unless $\alpha = \beta$ and $a_\alpha + b_\beta = 0$, in which case the value is $> \alpha$. Thus in any case

$$v(f+g) \geqslant \min\{v(f), v(g)\}.$$

It follows that $k\Gamma$ is an integral domain; if K is its field of fractions, and v is extended to K, then K is a valuated field with precise value group Γ.

Exercises

1. Give a direct proof of Corollary 7.2.
2. Let Γ be an abelian ordered group and consider the set of all 'power series' $\Sigma a_\alpha x^\alpha$ for which the *support*, i.e. the set of elements α with non-zero coefficient, is well ordered. Verify that these series form a field containing $k\Gamma$ on which a valuation can be defined as in the text, and that this field is complete.
3. Consider the special case $\Gamma = \mathbf{Z}$ of Exercise 2; thus we have the field $E = k((y))$ of all formal Laurent series in an indeterminate y. Now form $F = E((x))$, with a valuation w over E equal to the least element in the support on $\langle x \rangle$. Secondly, define a valuation ϕ on F with values in $\mathbf{Z}^2$ (ordered lexicographically), by

$$\phi(f) = \min\{(i, j) \mid a_{ij} \neq 0, \text{ where } f = \Sigma a_{ij} x^i y^j\}.$$

   Verify that ϕ is a valuation on F and further, show that $w(f) > 0 \Rightarrow \phi(f) > 0$, but that w and ϕ are not equivalent. Show that w can be decomposed as $\pi\phi$, where π is the projection of $\mathbf{Z}^2$ on its first factor.
4. Let K be an ordered field (i.e. a field with a total ordering '>' which is preserved by addition and by multiplication by elements that are > 0). Show that if F is any subfield of K, then the convex subring generated by F is a valuation ring V whose residue class field is an extension of F (more precisely, it can be shown that the residue class field lies between F and its completion as an ordered field). Verify that when F is the prime subfield, then the residue class field is archimedean ordered.
5. Let K be a field with a valuation v and residue class field k. Show that if k is archimedean ordered, then K has an ordering inducing the valuation v (as in Exercise 4). Obtain a generalization to the case where the residue class field is an arbitrary ordered field.
6. Let K be a field with a valuation ring V. If $\mathfrak{p}$ is a minimal non-zero prime ideal of V, show that the localization $V_\mathfrak{p}$ is a valuation corresponding to a rank 1 valuation of K.

2

Extensions

A central part of our topic is the study of extensions of rings of algebraic integers and of algebraic functions. In each case we are dealing with integral extensions of Dedekind domains, and we show here in section 2.4 how Dedekind domains can be characterized in terms of families of valuations satisfying the strong approximation theorem. The latter allows us to study the extensions locally, i.e. in terms of the different valuations, and this is done in sections 2.1–2.3. The information is put together in section 2.5 and is used in section 2.6 to describe two important invariants of an extension: the different and the discriminant.

2.1 GENERALITIES ON EXTENSIONS

Let $K \subseteq L$ be an extension of fields and suppose that w is a valuation on L whose restriction to K is v. We denote by Γ the precise value group of v, by V its valuation ring in K and by $\mathfrak{p}$ its maximal ideal; the residue class field $V/\mathfrak{p}$ will be written K_v. The corresponding objects for w are denoted by $\Delta, W, \mathfrak{P}, L_w$. Then we have

$$\mathfrak{p} = V \cap \mathfrak{P}, \qquad K_v = V/\mathfrak{p} = V/(V \cap \mathfrak{P}) \cong (V + \mathfrak{P})/\mathfrak{P} \subseteq W/\mathfrak{P} = L_w.$$

Thus we may regard the field L_w as an extension of K_v. In many cases this extension is finite and the degree of L_w over K_v, i.e. the dimension of L_w as K_v-space:

$$f = [L_w : K_v]$$

is called the *residue degree* of the extension L/K. Further, Γ is a subgroup of Δ and the index

$$e = (\Delta : \Gamma)$$

is called the *ramification index* of the extension L/K. The extension is said to be *ramified* if $e > 1$, *unramified* otherwise.

Let us consider some examples:

(i) $K = \mathbf{Q}$ with the 3-adic valuation.

(a) When $L = K(\sqrt{5})$, we have $K_v = \mathbf{F}_3, L_w = \mathbf{F}_3(\sqrt{5})$ and $e = 1$, $f = 2$. This follows because $x^2 - 5$ is irreducible over the 3-adic completion $\mathbf{Q}_3$ (we cannot even solve the congruence $x^2 \equiv 5 \pmod 3$ in $\mathbf{Z}$).

(b) The polynomial $x^2 - 7$ is reducible over $\mathbf{Q}_3$, for we have $\sqrt{7} = (1 + 2.3)^{1/2}$, and this binomial series converges in the 3-adic topology. Hence for $K(\sqrt{7})$ we have $e = f = 1$.

(c) The extension $K(\sqrt{3})$ is ramified, for the polynomial $x^2 - 3$ becomes x^2 over the residue class field $\mathbf{F}_3$; thus $e = 2, f = 1$.

(ii) $K = \mathbf{C}(t)$ with the t-adic valuation (defined by t as prime element). The residue class field is $\mathbf{C}$ which is algebraically closed, so $f = 1$ in all algebraic extensions.

(a) $K = K(\sqrt{(1 + t^2)})$. The polynomial $x^2 - (t^2 + 1)$ is irreducible, but over the residue class field it becomes $x^2 - 1$, which is reducible. In the completion we have

$$\sqrt{(1 + t^2)} = (1 + t^2)^{1/2}$$

which converges in the t-adic topology; $e = 1$.

(b) $L = K(\sqrt{t})$; the polynomial $x^2 - t$ becomes x^2 over the residue class field. The extension is ramified: $e = 2$.

(iii) $K = \mathbf{R}(t)$ with the t-adic valuation, $L = K(\sqrt{(t^2 - 1)})$. Put $u = \sqrt{(t^2 - 1)}$; we have $u^2 = t^2 - 1$, hence in the residue class field L_w, $\bar{u}^2 + 1 = 0$, so L_w is a quadratic extension of $\mathbf{R} = K_v$ and $f = 2$, $e = 1$.

It is clear that under repeated extensions the degrees $[L : K]$, f and e are all multiplicative. Thus if $K \subseteq L \subseteq M$ and f_{LK}, e_{LK} are the residue degree and ramification index of L/K etc., then

$$[M : K] = [M : L][L : K], \qquad f_{MK} = f_{ML} f_{LK}, \qquad e_{MK} = e_{ML}\, e_{LK}.$$

Of these the first is well known (and easily proved, cf. e.g. 3.1, p. 64 of A.2), hence the second follows, while the third follows from the corresponding formula for groups. To obtain a precise relation between these numbers we need to compare the different completions, and this will be done in section 2.3; but we can already establish a basic inequality:

Theorem 1.1

Let L/K be an extension of valuated fields, where the valuation w of L extends that of K, v say. If L/K is finite, then

$$ef \leqslant [L : K], \tag{1}$$

and if v is trivial or principal, then so is w. If v is principal and K is complete, then equality holds in (1).

Proof

Write V, W for the valuation rings of v, w respectively and K_v, L_w for their respective residue class fields. If $u_1, \ldots, u_r \in W$ are such that their residues $\bar{u}_1, \ldots, \bar{u}_r$ are linearly independent over the residue class field K_v, then for any $\alpha_1, \ldots, \alpha_r \in K$,

$$w(\alpha_1 u_1 + \ldots + \alpha_r u_r) = \min\{v(\alpha_1), \ldots, v(\alpha_r)\}. \tag{2}$$

Thus in particular, the u_i must be linearly independent over K and the value of any K-linear combination of the u's lies in Γ, the value group of v. To establish equation (2), we number the u's so that

$$v(\alpha_1) \leqslant v(\alpha_i) \quad (i > 1).$$

Since $\bar{u}_i \neq 0$, we have $w(u_i) = 0$, so the left-hand side of (2) has value at least $v(\alpha_1)$. If the value is greater, then after dividing by α_1 and passing to the residue class field, we have

$$\bar{u}_1 + (\overline{\alpha_2/\alpha_1})\bar{u}_2 + \ldots + (\overline{\alpha_r/\alpha_1})\bar{u}_r = 0,$$

which contradicts the linear independence of the $\bar{u}$'s. Hence (2) follows.

Now denote by Γ, Δ the value groups of v, w and take $\pi_1, \ldots, \pi_s \in L$ such that the values $w(\pi_j)$ are incongruent mod Γ. We claim that the rs elements $u_i\pi_j$ are linearly independent over K. For, given a relation

$$\sum a_{ij} u_i \pi_j = 0, \quad \text{where } a_{ij} \in K,$$

on writing $a_j = \sum a_{ij} u_i$, we have

$$\sum a_j \pi_j = 0.$$

Hence $w(a_h\pi_h) = w(a_k\pi_k)$ for some $h, k,\ h \neq k$. Thus

$$w(a_h/a_k) = w(\pi_k/\pi_h);$$

here the left-hand side is in Γ, because $w(a_i) \in \Gamma$ (by the remark after (2)), but the right-hand side is not in Γ, and this is a contradiction. Therefore all the a_j vanish: $\Sigma a_{ij} u_i = 0$. By what has been shown, all the a_{ij} vanish, so the $u_i\pi_j$ are linearly independent over K. It follows that $rs \leqslant [L:K]$. Hence e, f are finite and on taking $r = f$, $s = e$ we obtain (1).

If v is trivial, then Δ is finite of order e, but the only finite ordered group is the trivial group, so w must be trivial and $e = 1$. If v is principal, we may take $\Gamma = \mathbf{Z}$. Then $e\Delta \subseteq \mathbf{Z}$ and every non-trivial subgroup of $\mathbf{Z}$ is isomorphic to $\mathbf{Z}$, so $\Delta \cong \mathbf{Z}$, hence w is then also principal.

If v is principal and K is complete, then so is L, by Lemma 1.3.2, and w is again principal, as we have just seen. Taking uniformizers π, Π for v, w respectively, we can write every element of L uniquely as

$$x = \sum_{-N}^{\infty} \alpha_i \Pi^i, \tag{3}$$

where α_i lies in a transversal of the residue class field L_w of L in W. Here we can replace the powers Π^i by $\pi^i, \pi^i\Pi, \ldots, \pi^i\Pi^{e-1}$ and instead of a transversal of L_w take elements $u_1, \ldots, u_f$ in W such that $\bar{u}_1, \ldots, \bar{u}_f$ form a basis of L_w over K_v and then use $\Sigma\, \alpha_i u_i$, where α_i ranges over a transversal of K_v in V. Then (3) takes the form

$$x = \sum_{i=-N}^{\infty} \sum_{\mu\nu} a_{\mu\nu i} u_\mu \Pi^\nu \pi^i, \quad \text{where } a_{\mu\nu i} \in K.$$

This shows that L is spanned by the elements $u_\mu \Pi^\nu$ over K, so we find that $[L : K] \leqslant ef$ and equality in (1) follows.■

Without completeness we cannot expect equality in the expression (1), as the closer analysis in section 2.3 will show.

Any algebraic extension L/K can be written as a union of finite extensions, and applying Theorem 1.1 to these extensions we obtain the following conditions on the value group and the residue class field:

Corollary 1.2

If L/K is an algebraic extension and K has a valuation v with extension w to L, and corresponding value groups Γ, Δ and residue class fields K_v, L_w then Δ/Γ is a torsion group and L_w/K_v is algebraic.■

Exercises

1. Consider the valuation on $\mathbf{Q}(\sqrt{2})$ with prime element π, where π is one of $\sqrt{2}$, $1+\sqrt{2}$, $1-\sqrt{2}$, 5. In each case find the restriction of the valuation to $\mathbf{Q}$ and calculate e and f for the extension $\mathbf{Q}(\sqrt{2})/\mathbf{Q}$.
2. Let p, q be prime numbers and $E = \mathbf{Q}(\sqrt{q})$. Describe an extension of the p-adic valuation on $\mathbf{Q}$ to E and find e, f, distinguishing the cases $p \neq q$ and $p = q$. Do the same with q replaced by a square-free number, which may be composite.
3. Show that an algebraically closed field cannot have a discrete valuation (except the trivial one).

2.2 EXTENSIONS OF COMPLETE FIELDS

Let K be a valuated field and L an extension; we shall now examine ways of extending a valuation on K to L. By induction it will be enough to look at simple extensions, $L = K(\alpha)$, and we shall treat the cases where α is algebraic or transcendental separately. Thus our first task is to extend a valuation v on K to the rational function field $K(t)$, where t is transcendental over K. We shall give a simple construction of a valuation w on $K(t)$ which extends v (but shall not be concerned with finding *all* such extensions).

Let us define w on the polynomial ring $K[t]$ by putting, for

$$f = a_0 + a_1 t + \ldots + a_n t^n,$$

$$w(f) = \min\{v(a_0), \ldots, v(a_n)\}. \tag{1}$$

It is clear that this reduces to v on K, and moreover satisfies

$$w(f+g) \geqslant \min\{w(f), w(g)\};$$

further, the argument used to prove Gauss's lemma shows that $w(fg) = w(f) + w(g)$ (cf. Lemma 3.1.2). Hence we have a valuation w on $K[t]$; it extends in a unique way to a valuation on $K(t)$, still denoted w, by the rule

$$w(f/g) = w(f) - w(g). \tag{2}$$

The proof is the analogue of Proposition 1.1.1 for general valuations. The valuation thus defined is called the *Gaussian extension* of v to $K(t)$. Clearly v and w have the same value groups. Further, $w(t) = 0$ by (1), and the image $\bar{t}$ of t in the residue class field L_w of $L = K(t)$ is transcendental over K_v, for if $\bar{t}$ satisfies a polynomial equation over K_v, then there is a polynomial

$$f = \alpha_0 + \alpha_1 t + \ldots + \alpha_r t^r \quad (\alpha_i \in K),$$

such that $v(\alpha_i) \geqslant 0$, with equality for at least one i, such that $\bar{f} = \Sigma \bar{\alpha}_i \bar{t}^i = 0$. But this means that $w(f) > 0$, whereas by (1), $w(f) = 0$. This contradiction shows that no such f can exist, so $\bar{t}$ is transcendental over K_v and L_w is just the rational function field $K_v(\bar{t})$. We can sum up our conclusion as follows:

Theorem 2.1
Let K be any field with a general valuation v and $L = K(t)$ a purely transcendental extension. Then the Gaussian extension of v to L is a valuation on L, with the same value group as v and with residue class field $K_v(\bar{t})$, a purely transcendental extension of the residue class field K_v of K. ■

Usually there are other extensions of v to $K(t)$ which, however, will not be needed here (cf. Exercise 2).

Consider next an algebraic extension $K(\alpha)/K$. This is of finite degree, so if K is complete under v, there can be at most one extension to $K(\alpha)$, by Lemma 1.3.2. In fact there always is an extension, as we shall soon see. The original proof used Hensel's lemma, which tells us how to lift factorizations of a polynomial from the residue class field of K to K itself (cf. the remarks after Proposition 2.6). Another method, using Chevalley's extension lemma

(Lemma 8.4.3, p. 290 of A.2), applies quite generally to any extension, algebraic or not; moreover it does not assume completeness and applies to general valuations. Here we shall use a third method (taken from lectures by Dieudonné) which is very simple and though not applicable to the general case, is enough to deal with rank 1 valuations, which is all we shall need. We digress briefly to describe the notion of integral element and integral closure, which will play an important role later in this book.

Let K be a field and R a subring of K. An element c of K is said to be *integral over* R if it satisfies a monic equation over R:

$$c^n + a_1c^{n-1} + \ldots + a_n = 0, \quad \text{where } a_i \in R. \tag{3}$$

If every element of K integral over R lies in R, then R is said to be *integrally closed* in K. An integral domain is called *integrally closed* if it is integrally closed in its field of fractions.

The set of all elements of K integral over R is called the *integral closure* of R in K; it always includes R itself because any c in R satisfies the equation $x - c = 0$ and so is integral over R. Moreover, the integral closure of R is a subring of K; to prove this we put the definition in an alternative form:

Lemma 2.2

Let R be a subring of a field K. An element c of K is integral over R if and only if there is a non-zero finitely generated R-submodule M of K such that $cM \subseteq M$.

Proof

If c is integral over R, satisfying (3) say, then the R-module generated by $1, c, c^2, \ldots, c^{n-1}$ satisfies the given condition. Conversely, if M is a non-zero R-submodule of K such that $cM \subseteq M$ and M is generated by $u_1, \ldots, u_n$, then we have

$$cu_i = \sum a_{ij}u_j, \quad \text{where } a_{ij} \in R.$$

This is a homogeneous linear system of equations and not all the u_i vanish, because $M \neq 0$. Therefore its determinant is zero; the system may be written as

$$\sum (c\delta_{ij} - a_{ij})u_j = 0,$$

and we therefore have $\det(cI - A) = 0$, where $A = (a_{ij})$. On expanding this expression we obtain a monic equation for c over R, and this shows c to be integral over R.■

To show that the integral closure of R is a subring we have to show that the sum and product of elements integral over R are again integral over R. Given elements c_1, c_2 integral over R, say $c_iM_i \subseteq M_i$, for a non-zero finitely generated

R-submodule M_i of $K(i = 1, 2)$, we have $(c_1 + c_2)\, M \subseteq M$, where $M = M_1 + M_2$, and $c_1 c_2 N \subseteq N$, where

$$N = M_1 M_2 = \left\{ \sum x_i y_i \mid x_i \in M_1,\, y_i \in M_2 \right\}.$$

Clearly M, N are again finitely generated and non-zero, and by the lemma, this shows $c_1 + c_2$ and $c_1 c_2$ to be integral over R. Thus we have:

Corollary 2.3
For any subring R of a field K, the set of all elements integral over R forms a subring of K containing R. ■

Clearly R is integrally closed in K if and only if it equals its integral closure. It is an interesting fact that the integral closure of R in K may also be described as the intersection of all (general) valuation rings in K containing R, but we shall not need this result here (cf. e.g. Theorem 8.4.4, p. 291 of A.2). All we need is the fact that any valuation ring V in a field K is integrally closed, and this is easily seen: if c is integral over V but not in V, then c^{-1} lies in the maximal ideal $\mathfrak{m}$ of V and c satisfies (3) say, where $a_i \in V$. Hence

$$1 = -(a_1 c^{-1} + a_2 c^{-2} + \ldots + a_n c^{-n}) \in \mathfrak{m},$$

a contradiction.

We now return to the task in hand, of proving the existence of extensions of valuations in the algebraic case:

Proposition 2.4
Let L/K be a field extension of finite degree. Then any rank 1 valuation on K has an extension to L, again of rank 1.

Proof
Let v be the given valuation on K; as we know, v is characterized by its valuation ring V, which is a maximal proper subring of K, by Corollary 1.7.3. Hence K is the field of fractions of V. Let I be the integral closure of V in L; then the field of fractions of I is L. For any element γ of L is algebraic over K, hence it satisfies an equation with coefficients in K. On multiplying by a common denominator we obtain an equation over V:

$$a_0 x^n + a_1 x^{n-1} + \ldots + a_n = 0, \text{ where } a_i \in V,\ a_0 \neq 0.$$

It follows that $a_0\gamma$ is integral over V, for it satisfies the equation

$$y^n + a_1 y^{n-1} + a_2 a_0 y^{n-2} + \ldots + a_n a_0^{n-1} = 0.$$

Therefore $a_0\gamma \in I$ and this shows the field of fractions of I to be L.

Now consider the family $\mathscr{F}$ of all subrings R of L such that

$$I \subseteq R \subset L, \qquad R \cap K = V.$$

Since V is integrally closed in K, we have $I \cap K = V$, so the family $\mathcal{F}$ contains I. It is clearly inductive, and hence by Zorn's lemma, it has a maximal member W, say. We have to show that W is a maximal subring of L. In the first place, W is not a field, for the only field containing I is L and $L \cap K = K \neq V$. Suppose now that there is a ring R such that $W \subset R \subset L$ and take $c \in R \backslash W$. Then $W[c] \supset W$; by the maximality of W, $W[c] \cap K \supset V$, and hence $W[c] \supseteq K$, by the maximality of V in K. Thus $L \supseteq W[c] \supseteq K$; since L is of finite degree over K, it follows that $W[c]$ is a field and since $W[c] \supseteq I$, we have $W[c] = L$. This shows W to be maximal in L, as claimed. By Corollary 1.2 the quotient of the value groups is a torsion group, hence the two value groups have the same rank (Proposition 1.6.8), so the extension has again rank 1.■

Although this proof is anything but constructive, in the complete case we can easily determine the extension explicitly, thanks to its uniqueness:

Theorem 2.5

Let K be a field which is complete under a rank 1 valuation v and let L be a finite extension of K of degree n. Then v has a unique extension w to L, given by

$$w(\alpha) = \frac{1}{n} \cdot v(N_{L/K}(\alpha)), \text{ for all } \alpha \in L, \tag{4}$$

where N denotes the norm.

Proof

An extension exists by Proposition 2.4 and is unique by Theorem 1.3.3. To find its value we take an extension E of L such that E/K is normal. Then v has a unique extension w to E and for any automorphism σ of E/K and any $\alpha \in E$ there is an isomorphism $K(\alpha) \cong K(\alpha^\sigma)$ with α corresponding to α^σ in the isomorphism. By uniqueness it follows that $w(\alpha) = w(\alpha^\sigma)$. Now the norm of α from L to K is the product of the conjugates (which lie in E), $N_{L/K}(\alpha) = \alpha_1 \ldots \alpha_n$, where $\alpha \mapsto \alpha_i$ under a K-automorphism. Hence

$$v(N_{L/K}(\alpha)) = w(\alpha_1 \ldots \alpha_n) = \sum w(\alpha_i) = nw(\alpha),$$

and now (4) follows.■

We note that L/K has not been assumed separable here. The result remains true even in the archimedean case: by Ostrowski's theorem (Theorem 1.3.5) K must be $\mathbf{R}$ or $\mathbf{C}$ and so $[L : K]$ can only equal 1 or 2. In the only non-trivial case, $K = \mathbf{R}$, $L = \mathbf{C}$ and the norm on $\mathbf{C}$ is given by

$$\|\alpha\| = |\alpha\bar{\alpha}|^{1/2}.$$

Theorem 2.5 has the following interesting consequence:

Proposition 2.6
Let K be a field which is complete under a rank 1 valuation v and let $f = x^n + a_1 x^{n-1} + \ldots + a_n$ be an irreducible polynomial over K such that $v(a_n) \geqslant 0$. Then $v(a_i) \geqslant 0$ for $i = 1, \ldots, n$.

Proof
Take a minimal splitting field E of f over K and let $\alpha_1, \ldots, \alpha_n$ be the roots of $f = 0$ (not necessarily distinct). By Theorem 2.5, v has an extension w to E and

$$w(\alpha_i) = (1/n)v(a_n) \geqslant 0.$$

Thus the α_i are w-integers and hence the a_j as elementary symmetric functions of the α_i are v-integers, i.e. $v(a_j) \geqslant 0$, as claimed. ■

Using this result one can show that in the situation of Proposition 2.6, the image of f in the residue class field is a power of an irreducible polynomial. Hence any factorization of a polynomial over the residue class field into two coprime factors can be lifted to a factorization over K; this is just the assertion of Hensel's lemma (cf. e.g. Theorem 8.5.6, p. 299 of A.2). As mentioned earlier, this can also be proved directly and then used to provide another proof of Theorem 2.5.

We recall from standard field theory that a general field extension is obtained by taking a purely transcendental extension, followed by an algebraic extension (cf. e.g. Chapter 5 of A.3). Thus if we are given a valuated field K, we can form the rational function field $K(t)$ with the Gaussian extension. Repeating this process and applying Theorem 2.5 to the final algebraic extension, we have found an extension of a complete rank 1 valuation to any finitely generated extension.† It is not difficult to extend the argument to extensions that are not finitely generated, but we shall not pursue this line here. For us it will be more important to construct extensions of incomplete fields; this will be done in the next section.

Frequently we shall have to consider valuations that are trivial on a subfield. It will be convenient to describe a valuation on a field K whose restriction to a subfield k is trivial, as a valuation on K *over* k or a valuation on K/k. If v is a valuation on K/k with residue class field K_v, then it is clear that k is isomorphic to a subfield of K_v. There is a useful inequality relating the transcendence degrees of K/k and K_v/k; here the *transcendence degree* of L/K, written tr. deg. L/K, is the maximum number of elements of L that are algebraically independent over K.

Theorem 2.7
Let K be a field with a subfield k and let v be a valuation of K/k. Write V for the valuation ring, Γ for the value group and K_v for the residue class field of v.

† One would also need to show that when K is complete, then $K(t)$ with the Gaussian extension is again complete; but this will not be needed in what follows.

If tr. deg. K/k is finite, then

$$\text{tr. deg. } K_v/k + rk\,\Gamma \leqslant \text{tr. deg. } K/k. \tag{5}$$

Proof

We begin by showing that for any non-trivial valuation v on K/k the residue class field K_v has lower transcendence degree over k than K. Let $x_1, \ldots, x_n$ be a transcendence basis of K over k; writing $K_0 = k(x_1, \ldots, x_n)$, we have an algebraic extension K/K_0. On replacing x_i by x_i^{-1} if necessary we may assume that $v(x_i) \geqslant 0$, so that the valuation ring V contains $k[x_1, \ldots, x_n]$. Let us denote the residue class map by a bar (as before). The residue class field K_v is algebraic over $k(\bar{x}_1, \ldots, \bar{x}_n)$, for, given $\alpha \in K_v$, suppose that $\alpha = \bar{a}$, where $a \in K$ and take an equation for a over K_0:

$$c_0 a^n + c_1 a^{n-1} + \ldots + c_n = 0, \quad c_i \in K_0.$$

If we divide by the c_i of least value and pass to the residue class field, we obtain an equation for $\alpha = \bar{a}$ over $k(\bar{x}_1, \ldots, \bar{x}_n)$ in which a coefficient is 1, so we have a non-trivial equation for α and this shows K_v to be algebraic over $k(\bar{x}_1, \ldots, \bar{x}_n)$; it remains to show that $\bar{x}_1, \ldots, \bar{x}_n$ are algebraically dependent over k. If v is trivial on K_0, then it is trivial on the algebraic extension K (by Corollary 1.2). This is not so, hence v is non-trivial on K_0 and so $V_0 = V \cap K_0$ is a proper subring of K_0. The maximal ideal of V_0 contains a non-zero polynomial in the x's:

$$f = f(x_1, \ldots, x_n) \text{ say.}$$

We have

$$\bar{f} = f(\bar{x}_1, \ldots, \bar{x}_n) = 0,$$

thus $\bar{x}_1, \ldots, \bar{x}_n$ are algebraically dependent, so K_v has transcendence degree less than n over k.

To prove (5), take $r \leqslant \text{rk}\,\Gamma$; then there is a chain of r convex subgroups in Γ and hence a tower of r valuation rings above V:

$$K \supset V_1 \supset V_2 \supset \ldots \supset V_r = V.$$

Hence the place of K in K_v can be decomposed into r places; each lowers the transcendence degree by at least 1, and so we have

$$\text{tr. deg. } K_v/k \leqslant \text{tr. deg. } K/k - r. \tag{6}$$

This gives an upper bound for r, therefore rk Γ is finite, and taking $r = rk\,\Gamma$, we obtain (5) from (6).■

We remark that the inequality (5) may well be strict (cf. Exercise 3).

The method of Theorem 2.7 also allows us to recognize when the residue class field is algebraic over k:

Corollary 2.8
Let K/k be a field extension and v a valuation on K/k with valuation ring V and residue class field K_v. Then K_v is algebraic over k if and only if V is minimal among valuation subrings containing k.

Proof
Suppose that V is not minimal, say $V \supset W \supset k$, where W is a valuation ring with valuation w. Then w corresponds to a non-trivial valuation on K_v/k, so K_v cannot be algebraic over k, by Corollary 1.2. Conversely, if K_v/k is transcendental, say $K_v \supseteq k(x) \supset k$, where x is transcendental over k, then we can take the Gaussian extension of the trivial valuation on k to $k(x)$ and extend this to K_v by Proposition 2.4. Composing this valuation with v, we obtain a valuation w on K to which v is subordinate, i.e. its valuation ring W satisfies $V \supset W \supset k$, so V is not minimal. ■

It is possible to define a finer invariant on Γ, the *rational rank*, as the maximum number of elements in Γ linearly independent over $\mathbf{Z}$. For the rational rank an inequality similar to (5) holds, but this will not be needed here (cf. Bourbaki (1972)).

Exercises

1. Let K be a field with a rank 1 valuation v. Define a function w on the rational function field $K(x)$ by putting $w(f/g) = w(f) - w(g)$ and for a polynomial $f = \Sigma a_i x^i$ writing $w(f) = \min\{v(a_i) + \lambda i\}$, where λ is a real number not commensurable with any value of v. Show that w is a valuation extending v; find its value group and residue class field, distinguishing the cases $\lambda > 0$ and $\lambda < 0$.
2. Let K be a field with a valuation v, valuation ring V, maximal ideal $\mathfrak{p}$ and let u be a polynomial in a variable x over V which is irreducible (mod $\mathfrak{p}$). Fix a real number α and for any $f \in K[x]$ write $f = \Sigma f_i u^i$, where the f_i are polynomials of degree less than u, and define $w(f) = \min\{v(f_i) + \alpha i\}$, where v denotes the Gaussian extension of v to $K(x)$, and extend w to $K(x)$ as in Exercise 1. Verify that w is a valuation on $K(x)$ to which v is subordinate. Compare the value groups of v and w for various choices of α.
3. Let k be a field of characteristic 0 and $K = k(x, y)$ a purely transcendental extension in two variables. Given $f \in K$, we can in $k((t))$ write $f(t, e^t) = t^\nu f_1(t)$, where $f_1(0) \neq 0$ and e^t is the exponential series. Show that $v(f) = \nu$ defines a valuation on K. Verify that v has rank 1 and residue class field k.
4. Show that any valuation on K/k is subordinate to one with a residue class field that is algebraic over k. (*Hint:* Show that the family of valuation rings contained in a given one is closed under intersection and use Zorn's lemma.)

5. Let K be a field which is complete under a principal valuation v. Given a polynomial

$$f = x^n + c_1x^{n-1} + \ldots + c_n$$

over K, which is irreducible over K, show that if $v(c_n) > 0$, then $v(c_i) > 0$ for $i = 1, \ldots, n$.

6. Let L/K be a finite extension of degree n; let v be a principal valuation on K with valuation ring V and residue class field K_v, denote by w an extension to L, with valuation ring W and residue class field L_w. Suppose further that K (and hence also L) is complete, and that L_w/K_v is separable. Show that W has a V-basis of the form $1, \alpha, \ldots, \alpha^{n-1}$, for some $\alpha \in W$. (*Hint:* Take $\beta \in W$ such that $\bar{\beta}$ generates L_w over K_v and if f is the minimal polynomial for β over K, choose $\alpha \in W$ such that $w(\alpha - \beta) > 0$ and $w(f(\alpha)) = 1$.)

7. Let L/K be a Galois extension with group G. Suppose that K has a valuation v with valuation ring V, maximal ideal $\mathfrak{m}$ and residue class field K_v; let w be an extension of v to L, with corresponding entities W, $\mathfrak{n}$, L_w and define

$$G_0 = \{\sigma \in G \mid \alpha^\sigma \equiv \alpha (\text{mod } \mathfrak{n}) \text{ for all } \alpha \in L\}.$$

Show that G_0 is normal in G and G/G_0 is the Galois group of L_w/K_v (G_0 is called the *inertia subgroup* of G).

8. With the notations as in Exercise 7, put

$$G_i = \{\sigma \in G \mid \alpha^\sigma \equiv \alpha (\text{mod } \mathfrak{n}^{i+1})\}$$

for $i = 1, 2, \ldots$; and assume that $L_w \cong K_v$, thus $f = 1$. Show that $G_0 = G$, G_0/G_1 is embedded in $K^\times$ and G_i/G_{i+1} is embedded in $K^+ (i > 0)$; deduce that G is soluble.

2.3 EXTENSIONS OF INCOMPLETE FIELDS

We next consider algebraic extensions of fields with a rank 1 valuation that may be incomplete. The valuation can again be extended, by Proposition 2.4, but now the extension need not be unique. To study this problem we shall need the tensor product of algebras. For any algebras A, B over a field k their *tensor product* $C = A \otimes B$ is defined by the property that the bilinear maps on $A \times B$ correspond to the linear maps of C (cf. 4.7 and 5.5 of A.2). Explicitly, if A, B have bases $\{u_i\}$, $\{v_j\}$ over k, then $A \otimes B$ has the basis $\{u_i \otimes v_j\}$: any element x of $A \otimes B$ can be uniquely written in any one of the forms

$$x = \Sigma\alpha_{ij}u_i \otimes v_j = \Sigma a_j \otimes v_j = \Sigma u_i \otimes b_i,$$

where $\alpha_{ij} \in k$, $a_j = \Sigma\alpha_{ij}u_i \in A$, $b_i = \Sigma\alpha_{ij}v_j \in B$. In particular, if $B = E$ is an extension field of k and $\{u_i\}$ is a k-basis of A, then it is also an E-basis for the E-algebra $A_E = A \otimes E$. If $[A : k] = n$, we have

$$[A_E : E] = n.$$

The tensor product $A \otimes B$, must be carefully distinguished from the direct product $A \times B$; the latter is the set of all pairs (a, b), where $a \in A$, $b \in B$, regarded as an algebra on which all operations are carried out componentwise: $(a, b) \pm (a', b') = (a \pm a', b \pm b')$. As additive group it may also be written in the form of a direct sum $A \oplus B$, but this can be misleading when the unit elements are considered: if A, B have unit elements e, f respectively, then $A \oplus B$ has the unit element $e + f$. Thus A, B are homomorphic images but not subrings of $A \oplus B$ (cf. 5.2 of A.2). Here we shall mainly be concerned with the direct product of a family of fields: from $K_1, \ldots, K_r$ we form the direct product $\Pi K_i = K_1 \times \ldots \times K_r$, where the operations are again performed componentwise.

Proposition 3.1
Let E, F be fields both containing k as subfield and assume that the extension F/k is separable of degree n. Then the tensor product has the form

$$E \otimes F \cong \Pi_1^r K_i, \tag{1}$$

where $K_1, \ldots, K_r$ are fields each containing an isomorphic copy of E and of F and generated by these copies.

The fields K_i are sometimes described as composites of E and F; formally a *composite* of E and F is a field containing copies of E and F and generated by these copies.

Proof
Since F/k is a separable extension, it is simple, say $F = k(\beta)$, where β has the minimal polynomial f of degree n over k. Then $F = k[x]/(f)$, i.e. we have the exact sequence

$$0 \to (f) \to k[x] \to F \to 0,$$

where (f) denotes the ideal of $k[x]$ generated by f. Since the terms are vector spaces over the field k, the sequence is split exact and so it remains exact on tensoring with E:

$$0 \to (f)_{E[x]} \to E[x] \to E \otimes F \to 0,$$

where $(f)_{E[x]}$ is the ideal generated by f in $E[x]$. Let $f = g_1 \ldots g_r$ be the decomposition of f into irreducible factors over E; since β was separable, there are no repeated factors. The quotient $K_i = E[x]/(g_i)$ is a field, because (g_i) is a maximal ideal in $E[x]$; therefore we have a surjective ring homomorphism:

$$\mu_i : E \otimes F \to K_i,$$

explicitly, for any $\phi \in E[x]$, μ_i maps $\phi(1 \otimes \beta)$ to $\phi(\beta_i)$, where β_i is a root of

$g_i(x)=0$. By composing these maps we obtain a homomorphism

$$E\otimes F\to K_1\times\ldots\times K_r. \tag{2}$$

Let $\phi\in E[x]$ and suppose that $\phi(\beta)(=\phi(1\otimes\beta))$ belongs to the kernel of the map (2). Then $g_i\mid\phi$ for $i=1,\ldots,r$, hence $f\mid\phi$ and so $\phi(\beta)=0$. Thus the map (2) is injective and by comparing dimensions (which on both sides are $\deg f$ over E) we see that it is bijective and hence an isomorphism. Further, μ_i restricted to E or F is a homomorphism of fields, hence injective, because the kernel is an ideal not containing 1, which in a field must be zero. This shows that each of E, F is embedded in K_i by μ_i and the images generate K_i (even as a ring) because μ_i is surjective.■

In the situation of Proposition 3.1 let us take $\alpha\in F$ and write ϕ, ψ_i for the characteristic polynomial of α over k and for that of the image of α in K_i over E. Then we have

$$\phi=\psi_1\ldots\psi_r. \tag{3}$$

For if $v_1,\ldots,v_t$ is a basis of F over k and $\alpha v_i=\Sigma a_{ij}v_j$, where $a_{ij}\in k$ then ϕ is given by $\phi(x)=\det(xI-A)$, where $A=(a_{ij})$. Clearly ϕ is also the characteristic polynomial of α as an element of $E\otimes F$ over E. We now change the basis in $E\otimes F$ to a basis adapted to the decomposition on the right of (1). On K_i, $\alpha\mu_i$ has the characteristic polynomial ψ_i and the matrix corresponding to A in the new basis is in block diagonal form, so we have $\phi=\psi_1\ldots\psi_r$, i.e. (3). In particular, comparing second and last coefficients, we find expressions for the trace and norm:

Corollary 3.2

If k, E, F, K_i are as in Proposition 3.1, then for any $\alpha\in F$,

$$N_{F/k}(\alpha)=\Pi N_{K_i/E}(\alpha),$$

$$\mathrm{Tr}_{F/k}(\alpha)=\Sigma\,\mathrm{Tr}_{K_i/E}(\alpha).■$$

We can now describe the extensions of an incomplete field:

Theorem 3.3

Let k be a field with a principal valuation v and let K/k be a separable extension of degree n. Then there are at most n extensions of v to K, say $w_1,\ldots,w_r$, where $r\leqslant n$. If f_i denotes the residue degree and e_i the ramification index of w_i, then

$$\Sigma e_i f_i=n.$$

Further, if the completion of k under v is $\hat{k}$ and that of K under w_i is K_i, then

$$K\otimes\hat{k}\cong K_1\times\ldots\times K_r. \tag{4}$$

Proof

By Proposition 3.1 $K \otimes \hat{k}$ has the form (4) for some fields K_i that are composites of K and $\hat{k}$. Hence v defined on $\hat{k}$ has a unique extension w_i to K_i (by Theorem 2.5), which by restriction defines w_i on K.

If $K \otimes \hat{k}$ is regarded as a topological ring, then K is a dense subfield, because its closure clearly contains $K\hat{k}$. Hence K is dense in each K_i, which is therefore the completion of K for w_i. The w_i are distinct, and hence inequivalent, because we can approximate each $(\alpha_1, \ldots, \alpha_r)$ of the right-hand side of (4) by elements of K, by Theorem 1.2.3. It remains to show that there are no other extensions. Let w be a further valuation extending v to K; by continuity it extends to $K \otimes \hat{k}$, and the latter is complete, as normed $\hat{k}$–space. If K is the completion of K relative to w, the $\hat{K}$ must equal one of the factors on the right of (4), say $\hat{K} \cong K_1$ and so we have $w = w_1$. So there are no extensions other than $w_1, \ldots, w_r$.

Let $[K_i : \hat{k}] = n_i$; by Theorem 1.1, $n_i = e_i f_i$ and by (4).

$$n = [K : k] = [K \otimes \hat{k} : \hat{k}] = \Sigma n_i,$$

hence $\Sigma e_i f_i = n$, as we had to show.■

We note that the result still holds for archimedean absolute values. In that case $\hat{k} = \mathbf{R}$ or $\mathbf{C}$ and if e.g. $K \otimes \mathbf{R} \cong K_1 \times \ldots \times K_r$, then $[K_i : \mathbf{R}] = 1$ or 2 according as K_i is real or complex, and we put $e_i = 1$ or 2 accordingly.

For Galois extensions the result just proved takes the following form.

Theorem 3.4

Let k be a field with a principal valuation v, let $\hat{k}$ be its completion and let K/k be a Galois extension of degree n. If $w_1, \ldots, w_r (r \leqslant n)$ are all the extensions of v to K which exist by Theorem 3.3, then $e_i = e$, $f_i = f$ and $[K : k] = efr$. Moreover, the valuations $w_1, \ldots, w_r$ are permuted transitively by the Galois group of K/k.

Proof

Consider an extension w_i of v and take $\sigma \in G$, where $G = \mathrm{Gal}(K/k)$. We can define a valuation σw_i on K by putting $(\sigma w_i)(\alpha) = w_i(\alpha^\sigma)$; it is easily checked that σw_i is indeed a valuation and that it reduces to v on k. To show that G acts transitively on the w_i it will be enough to find $\sigma \in G$ such that $\sigma w_1 = w_2$. Suppose that the two families of valuations $\{\sigma w_1\}$, $\{\tau w_2\}$ as $\sigma, \tau \in G$ are disjoint. By the approximation theorem (Theorem 1.2.3) there exists $a \in K$ such that $w_1(a^\sigma) > 0$, $w_2(a^\sigma - 1) > 0$ for all $\sigma \in G$. It follows that $w_2(a^\sigma) = 0$, so

$$v(N(a)) = \Sigma_\sigma w_2(a^\sigma) = 0$$

but we also have

$$v(N(a)) = \Sigma_\sigma w_1(a^\sigma) > 0$$

and this contradicts the previous equation. Hence $\sigma w_1 = \tau w_2$ for some σ, τ and so $w_2 = \tau^{-1}\sigma w_1$. This proves the last part. Now it follows that all the e_i are equal, to e say, and likewise $f_i = f$, and so $[K : k] = \Sigma_1^r ef = efr$.■

Exercises

1. Show that every composite of E and F occurs on the right-hand side of (1).
2. Show that without the assumption of separability in Proposition 3.1, (1) is replaced by $(E \otimes F)/N \cong \Pi K_i$, where N is the radical of the algebra $E \otimes F$. Find the modification of Theorem 3.3 necessary in this case.
3. Let K be a field of prime characteristic p with a rank 1 valuation v, and let E be a p-radical (=purely inseparable) extension of degree $[E : K] = q = p^r$. Show that v has a unique extension w to E, given by
$$w(\alpha) = (1/q)v(\alpha^q) \qquad (\alpha \in E).$$
4. Let K be a field with a valuation v. Given an element α algebraic over K, with minimal equation
$$x^n + a_1 x^{n-1} + \ldots + a_n = 0$$
and conjugates $\alpha_1 = \alpha, \alpha_2, \ldots, \alpha_n$ over K, show that for any extension w of v to a splitting field for α, the least value of $w(\alpha_i)$ is
$$\min\{v(a_1), v(a_2^{1/2}), \ldots, v(a_n^{1/n})\},$$
and that the number of conjugates for which this value is assumed is equal to the suffix of the first a_i for which $v(a_i^{1/i})$ has this minimum value.
5. Show that the field E of primitive 12th roots of 1 is of degree 4 over $\mathbf{Q}$, generated by a root of $x^4 - x^2 + 1 = 0$. Verify that the 11-adic valuation on $\mathbf{Q}$ has two extensions to E which are both unramified.

2.4 DEDEKIND DOMAINS AND THE STRONG APPROXIMATION THEOREM

As we have seen, divisibility in the ring of integers can be entirely described by valuations, and the same method clearly applies to any principal ideal domain, but in fact it holds for an even wider class of rings, of importance in what follows. The sets of valuations used will have to possess a property which is a strengthening of the approximation theorem (Theorem 1.2.3); this can be shown to be equivalent to the uniqueness of ideal multiplication and so lead to the class of Dedekind domains. Here we shall establish its equivalence with E. Noether's definition (Theorem 4.6 below).

Let K be a field with a family S of principal valuations defined on it. We may as well take the members of S to be non-trivial and pairwise inequivalent, thus the members of S are equivalence classes of valuations, also called *prime divisors* or simply *places*. We shall denote the members of S by small Gothic (or *Fraktur*) letters: $\mathfrak{p}, \mathfrak{q}$, ... and write $v_\mathfrak{p}, v_\mathfrak{q}$,... for the corresponding valuation. With each $\mathfrak{p} \in S$ we associate its valuation ring

$$\mathfrak{o}_\mathfrak{p} = \{x \in K \mid v_\mathfrak{p}(x) \geqslant 0\},$$

and we set

$$\mathfrak{o} = \bigcap \mathfrak{o}_\mathfrak{p}. \tag{1}$$

It follows that divisibility relative to $\mathfrak{o}$ is described by the rule

$$x \mid y \text{ if and only if } v_\mathfrak{p}(x) \leqslant v_\mathfrak{p}(y) \quad \text{for all } \mathfrak{p} \in S.$$

An element x of K is said to be *integral at* $\mathfrak{p}$ if $v_\mathfrak{p}(x) \geqslant 0$. We give some examples, which will play a role in the sequel.

1. The rational field $\mathbf{Q}$. Taking S to be the set of all prime numbers, with the p-adic valuation corresponding to p, we find that $\mathfrak{o} = \mathbf{Z}$.
2. $K = k(x)$, a rational function field, S the set of all valuations on K over k. If $a \in \mathfrak{o}$, then $v_\mathfrak{p}(a) \geqslant 0$ for all $\mathfrak{p} \in S$. Thus if

$$a = cp_1^{\alpha_1} \ldots p_r^{\alpha_r}, \quad c \in k^\times, p_i \text{ monic irreducible over } k,$$

   then for $a \in \mathfrak{o}$ we must have $\alpha_i \geqslant 0$, i.e. a is a polynomial in x. But as we saw, — deg a is also a valuation, so we also have deg $a \leqslant 0$, and so a reduces to c. Hence in this case $\mathfrak{o} = k$.
3. If we take $K = k(x)$ as in Example 2, but omit — deg from S, we find that $\mathfrak{o} = k[x]$.

Example 2 shows that some condition is needed to ensure that $\mathfrak{o}$ is not unreasonably small compared to K. This leads to the following conditions on the set S:

D.1 Each valuation is principal.

D.2 For any $x \in K$, $v_\mathfrak{p}(x) \geqslant 0$ for almost all $\mathfrak{p} \in S$.

D.3 Given $\mathfrak{p}, \mathfrak{p}' \in S$, $\mathfrak{p} \neq \mathfrak{p}'$ and $N > 0$, there exists $x \in K$ such that
$v_\mathfrak{p}(x - 1) > N$, $v_{\mathfrak{p}'}(x) > N$ and $v_q(x) \geqslant 0$ for all $\mathfrak{q} \neq \mathfrak{p}, \mathfrak{p}'$ in S.

A set S of valuations satisfying **D**.1–**D**.3 is said to possess the *strong approximation property*. By **D**.1, the valuations can be taken to be normalized. The first two conditions of **D**.3, relating to $\mathfrak{p}, \mathfrak{p}'$ can always be satisfied by the approximation theorem; the third one carries the sting. By demanding that the x to satisfy the condition can be taken to be integral it ensures that $\mathfrak{o}$ is large enough. For example, we easily see that **D**.1–**D**.3 entail that $\mathfrak{o}$ has K as its field of fractions. For take any $a \in K^\times$ and suppose that $v_\mathfrak{p}(a) \geqslant 0$ for all $\mathfrak{p}$ except $\mathfrak{p}_1, \ldots, \mathfrak{p}_n$ (by **D**.2). Pick $\mathfrak{q} \neq \mathfrak{p}_i$ and choose $b \in K$ such that $v_{\mathfrak{p}_i}(b) \geqslant - v_{\mathfrak{p}_i}(a)$, $v_\mathfrak{q}(b-1) \geqslant 0$ and $v_{\mathfrak{p}'}(b) \geqslant 0$ for $\mathfrak{p}' \neq \mathfrak{q}, \mathfrak{p}_i$. Then $b \in \mathfrak{o}$, $b \neq 0$ and $ab \in \mathfrak{o}$, say $ab = c$, therefore $a = cb^{-1}$.

The existence of such a b follows from a strengthening of Theorem 1.2.3, known as the *strong approximation theorem.*

Theorem 4.1

Let K be a field with a set S of valuations satisfying **D**.1–**D**.3. *Then for any distinct $\mathfrak{p}_1, \ldots, \mathfrak{p}_n \in S$, any $a_1, \ldots, a_n \in K$ and any $N > 0$ there exists $a \in K$ such that*

$$v_{\mathfrak{p}_i}(a - a_i) > N, \quad i = 1, \ldots, n, \tag{2}$$

$$v_{\mathfrak{q}}(a) \geqslant 0, \quad \mathfrak{q} \neq \mathfrak{p}_1, \ldots, \mathfrak{p}_n. \tag{3}$$

Proof

When S is finite, this is essentially Theorem 1.2.3 and there is nothing more to prove; so we may take S to be infinite and we may also assume that

$$v_{\mathfrak{q}}(a_i) \geqslant 0 \quad \text{for } \mathfrak{q} \neq \mathfrak{p}_1, \ldots, \mathfrak{p}_n.$$

For this can only fail to hold at finitely many $\mathfrak{q}$, which we can add to $\mathfrak{p}_1, \ldots, \mathfrak{p}_n$, putting the corresponding a_i equal to 0. We may also assume, without loss of generality, that $n > 1$. Let M be a positive constant; our aim is to construct b_1 integral such that

$$v_{\mathfrak{p}_1}(b_1 - 1) > M, \qquad v_{\mathfrak{p}_i}(b_1) > M \quad \text{for } i = 2, \ldots, n. \tag{4}$$

By **D**.3, for each $i = 2, \ldots, n$ there exists c_i integral at all $\mathfrak{q} \neq \mathfrak{p}_1, \mathfrak{p}_i$ such that

$$v_{\mathfrak{p}_1}(c_i - 1) > M, \quad v_{\mathfrak{p}_i}(c_i) > M.$$

Since $M > 0$, c_i is also integral at $\mathfrak{p}_1, \mathfrak{p}_i$ and so $c_i \in \mathfrak{o}$. Putting $b_1 = c_2 \ldots c_n$, we have $b_1 \in \mathfrak{o}$ and

$$v_{\mathfrak{p}_i}(b_1) = \Sigma_j v_{p_i}(c_j) > M.$$

Further, we have

$$\begin{aligned} b_1 - 1 &= c_2 \ldots c_n - 1 \\ &= (c_2 - 1)c_3 \ldots c_n + (c_3 - 1)c_4 \ldots c_n + c_n - 1, \end{aligned}$$

hence

$$v_{\mathfrak{p}_1}(b_1 - 1) \geqslant \min \{v_{\mathfrak{p}_1}(c_2 - 1), \ldots, v_{\mathfrak{p}_1}(c_n - 1)\} > M.$$

Thus b_1 satisfying (4) has been found. If we define $b_2, \ldots, b_n$ similarly and then put $a = \Sigma a_i b_i$, we find that

$$v_{\mathfrak{p}_1}(a - a_1) = v_{\mathfrak{p}_1}(a_1(b_1 - 1) + \Sigma_2^n a_i b_i) \geqslant \min_i \{v_{\mathfrak{p}_1}(a_i) + M\},$$

and it follows that $v_{\mathfrak{p}_1}(a - a_1) > N$, provided we choose M to satisfy

$$M > N - \min_{ij} \{v_{\mathfrak{p}_j}(a_i)\}.$$

Similarly $v_{\mathfrak{p}_i}(a - a_i) > N$ and for $\mathfrak{q} \neq \mathfrak{p}_i$,

$$v_{\mathfrak{q}}(a) \geqslant \min\{v_{\mathfrak{q}}(a_i) + v_{\mathfrak{q}}(b_i)\} \geqslant 0,$$

so a satisfies all the conditions.■

In order to apply this result, we shall need the notion of an invertible ideal in an integral domain, and its relation to prime divisors. We shall describe the general situation, leaving aside **D**.3 for the moment.

For any subring $\mathfrak{o}$ of a field K we define a *fractional ideal* of $\mathfrak{o}$ as an $\mathfrak{o}$-module $\mathfrak{A}$ in K such that

$$u\mathfrak{o} \subseteq \mathfrak{A} \subseteq v\mathfrak{o} \quad \text{for some } u, v \in K^{\times}.$$

An ordinary ideal $\mathfrak{A}$ of $\mathfrak{o}$ is a fractional ideal precisely if it is non-zero; this is also called an *integral* ideal. Clearly a fractional ideal is integral if and only if it is contained in $\mathfrak{o}$.

The usual ideal multiplication can be carried out for fractional ideals, defining $\mathfrak{A}_1\mathfrak{A}_2$ as

$$\mathfrak{A}_1\mathfrak{A}_2 = \{\Sigma x_\nu y_\nu \mid x_\nu \in \mathfrak{A}_1, y_\nu \in \mathfrak{A}_2\}.$$

If $u_i\mathfrak{o} \subseteq \mathfrak{A}_i \subseteq v_i\mathfrak{o}$, then $u_1u_2\mathfrak{o} \subseteq \mathfrak{A}_1\mathfrak{A}_2 \subseteq v_1v_2\mathfrak{o}$; hence the product is again a fractional ideal. This multiplication is clearly associative, with $\mathfrak{o}$ as unit element, so the set F of all fractional ideals is a monoid. Moreover, there is a generalized inverse:

$$(\mathfrak{o} : \mathfrak{A}) = \{x \in \mathfrak{o} \mid x\mathfrak{A} \subseteq \mathfrak{o}\}$$

If $u\mathfrak{o} \subseteq \mathfrak{A} \subseteq v\mathfrak{o}$, then $v^{-1}\mathfrak{o} \subseteq (\mathfrak{o} : \mathfrak{A}) \subseteq u^{-1}\mathfrak{o}$, and for any $c \in \mathfrak{o}$ we have

$$x\mathfrak{A} \subseteq \mathfrak{o} \Rightarrow cx\mathfrak{A} \subseteq c\mathfrak{o} \subseteq \mathfrak{o},$$

hence $(\mathfrak{o} : \mathfrak{A})$ is again a fractional ideal. Further, we have

$$(\mathfrak{o} : \mathfrak{A})\mathfrak{A} \subseteq \mathfrak{o}, \tag{5}$$

as is easily verified. If equality holds in (5), we also write $\mathfrak{A}^{-1}$ in place of $(\mathfrak{o} : \mathfrak{A})$ and call $\mathfrak{A}$ *invertible*. For example, any non-zero principal ideal $a\mathfrak{o}$ is invertible: $(a\mathfrak{o})^{-1} = a^{-1}\mathfrak{o}$.

Suppose now that $\mathfrak{o}$ is the ring of integers for a set S of places on K. We define the *divisor group* D of K with respect of S as the free abelian group on S as generating set. The typical element is written.

$$\mathfrak{a} = \prod \mathfrak{p}^{\alpha_\mathfrak{p}},$$

where the $\alpha_\mathfrak{p}$ are integers, almost all zero, and $\mathfrak{a}$ is called a *divisor*. From this point of view the elements of S are just the 'prime divisors'. Our aim will be

to explore the relations between D and the monoid F of all fractional ideals. We define a mapping $\phi : F \to D$ by the rule:
for any $\mathfrak{A} \in F$ we put

$$v_\mathfrak{p}(\mathfrak{A}) = \min \{v_\mathfrak{p}(x) \mid x \in \mathfrak{A}\}. \tag{6}$$

If $u\mathfrak{o} \subseteq \mathfrak{A} \subseteq v\mathfrak{o}$, then clearly $v_\mathfrak{p}(u) \geqslant v_\mathfrak{p}(\mathfrak{A}) \geqslant v_\mathfrak{p}(v)$ for all $\mathfrak{p}$; this shows that $v_\mathfrak{p}(\mathfrak{A}) = 0$ for almost all $\mathfrak{p}$. We can therefore define

$$\mathfrak{A}\phi = \prod \mathfrak{p}^{v_\mathfrak{p}(\mathfrak{A})}. \tag{7}$$

We remark that ϕ is a homomorphism. For if $c \in \mathfrak{A}\mathfrak{B}$, say $c = \Sigma a_i b_i$, ($a_i \in \mathfrak{A}, b_i \in \mathfrak{B}$) then

$$v_\mathfrak{p}(c) \geqslant \min_i \{v_\mathfrak{p}(a_i) + v_\mathfrak{p}(b_i)\} \geqslant v_\mathfrak{p}(\mathfrak{A}) + v_\mathfrak{p}(\mathfrak{B}),$$

therefore $v_\mathfrak{p}(\mathfrak{A}\mathfrak{B}) \geqslant v_\mathfrak{p}(\mathfrak{A})v_\mathfrak{p}(\mathfrak{B})$, and here equality holds, as we see by taking $c = ab$, where a, b are chosen in $\mathfrak{A}, \mathfrak{B}$ so as to attain the minimum in (6). Hence

$$v_\mathfrak{p}(\mathfrak{A}\mathfrak{B}) = v_\mathfrak{p}(\mathfrak{A}) + v_\mathfrak{p}(\mathfrak{B}),$$

and it follows that (7) is a homomorphism.

In general there is not much more that can be said about ϕ. We shall now impose the strong approximation property on S, so that Theorem 4.1 holds; this will allow us to conclude that ϕ is an isomorphism, and hence that every fractional ideal is invertible, so that F is a group. We need two lemmas:

Lemma 4.2
Let K be a field and S a family of places with the strong approximation property. Given any finite subset $\{\mathfrak{p}_1, \ldots, \mathfrak{p}_n\}$ of S and any integers $\alpha_1, \ldots, \alpha_n$, there exists $a \in K$ such that

$$v_{\mathfrak{p}_i}(a) = \alpha_i \quad (i = 1, \ldots, n), \qquad v_\mathfrak{q}(a) \geqslant 0 \quad \textit{for } \mathfrak{q} \neq \mathfrak{p}_i \quad (i = 1, \ldots, n). \tag{8}$$

Proof
Let $a_i \in K$ be such that $v_{\mathfrak{p}_i}(a_i) = \alpha_i (i = 1, \ldots, n)$; such a_i exists because $v_{\mathfrak{p}_i}$ is normalized. By Theorem 4.1 there exists $a \in K$ such that

$$v_{\mathfrak{p}_i}(a - a_i) > \alpha_i, \qquad v_\mathfrak{q}(a) \geqslant 0 \quad \text{for } \mathfrak{q} \neq \mathfrak{p}_i.$$

Hence $v_{\mathfrak{p}_i}(a) \geqslant \min \{v_{\mathfrak{p}_i}(a_i), v_{\mathfrak{p}_i}(a - a_i)\}$, and here equality holds, because

$$v_{\mathfrak{p}_i}(a_i) = \alpha_i < v_{\mathfrak{p}_i}(a - a_i);$$

thus (8) is satisfied.■

Lemma 4.3
Let K be a field and S a family of places with the strong approximation property, and let $\mathfrak{o} = \cap \{\mathfrak{o}_\mathfrak{p} \mid \mathfrak{p} \in S\}$ be the associated ring of integers. Given a fractional ideal $\mathfrak{A}$ in K, if $v_\mathfrak{p}(\mathfrak{A})$ is defined by (6), then for all $x \in K$,

$$x \in \mathfrak{A} \Leftrightarrow v_\mathfrak{p}(x) \geqslant v_\mathfrak{p}(\mathfrak{A}) \quad \textit{for all } \mathfrak{p} \in S. \tag{9}$$

Proof
The implication $\Rightarrow$ holds by definition; to prove that $\Leftarrow$ holds, let us fix x in $\mathfrak{A}$ and replace $\mathfrak{A}$ by $x^{-1}\mathfrak{A}$; then we have to show:

$$v_\mathfrak{p}(\mathfrak{A}) \leqslant 0 \Rightarrow 1 \in \mathfrak{A}.$$

If we replace $\mathfrak{A}$ by $\mathfrak{A} \cap \mathfrak{o}$, then $\mathfrak{A} \subseteq \mathfrak{o}$ and the hypothesis becomes $v_\mathfrak{p}(\mathfrak{A}) = 0$. Take $c \in \mathfrak{A}^\times$; if c is a unit, then $1 = cc^{-1} \in \mathfrak{A}$. Otherwise $v_\mathfrak{p}(c) \neq 0$ for only finitely many places, say $\mathfrak{p} = \mathfrak{p}_1, \ldots, \mathfrak{p}_n$. Writing v_i for $v_{\mathfrak{p}_i}$, let us take $a_i \in \mathfrak{A}$ such that $v_i(a_i) = 0$. Now fix j, $1 \leqslant j \leqslant n$, and by Theorem 4.1 choose $b_j \in K$ such that

$$v_j(a_j^{-1} - b_j) \geqslant v_j(c), \qquad v_\mathfrak{q}(b_j) \geqslant v_\mathfrak{q}(c) \quad \text{for } \mathfrak{q} \neq \mathfrak{p}_j.$$

Since $v_\mathfrak{q}(c) = 0$ at almost all places, this is possible, and in fact $b_j \in \mathfrak{o}$ because $c \in \mathfrak{o}$ and $v_j(a_j) = 0$. If we carry out this construction for $j = 1, \ldots, n$ and put $a = \Sigma a_i b_i$, we find that $a \in \mathfrak{A}$, because $a_i \in \mathfrak{A}$ and $b_i \in \mathfrak{o}$. Further, by the choice of b_i we have

$$v_j(1 - a) = v_j((1 - a_j b_j) - \sum_{i \neq j} a_i b_i) \geqslant v_j(c), \qquad j = 1, \ldots, n,$$

and $1 - a \in \mathfrak{o}$, so for $\mathfrak{q} \neq \mathfrak{p}_1, \ldots, \mathfrak{p}_n$,

$$v_\mathfrak{q}(1 - a) \geqslant 0 = v_\mathfrak{q}(c).$$

Hence $c^{-1}(1 - a) = d \in \mathfrak{o}$ and $1 = a + cd \in \mathfrak{A}$, as claimed.■

In order to show that ϕ is an isomorphism we shall define a mapping $\psi : D \to F$ which will turn out to be the inverse of ϕ. It is defined by the rule,

$$\text{if } \mathfrak{a} = \Pi \mathfrak{p}^{\alpha_\mathfrak{p}}, \quad \text{then} \quad \mathfrak{a}\psi = \{x \in K \mid v_\mathfrak{p}(x) \geqslant \alpha_\mathfrak{p} \quad \text{for all } \mathfrak{p} \in S\}.$$

It is clear that $\mathfrak{a}\psi$ is an $\mathfrak{o}$-module; to show that it is a fractional ideal, let $\mathfrak{p}_1, \ldots, \mathfrak{p}_r$ be the places for which $\alpha_\mathfrak{p} \neq 0$, take $u \in K$ such that $v_\mathfrak{p}(u) = \alpha_\mathfrak{p}$ for $\mathfrak{p} = \mathfrak{p}_1, \ldots, \mathfrak{p}_r$ and $v_\mathfrak{q}(u) \geqslant 0$ for $\mathfrak{q} \neq \mathfrak{p}_1, \ldots, \mathfrak{p}_r$ and take $v \in K$ such that $v_\mathfrak{p}(v) = -\alpha_\mathfrak{p}$ for $\mathfrak{p} = \mathfrak{p}_1, \ldots, \mathfrak{p}_r$ and $v_\mathfrak{q}(v) \geqslant 0$ for $\mathfrak{q} \neq \mathfrak{p}_1, \ldots, \mathfrak{p}_r$. Then it is clear that $u\mathfrak{o} \subseteq \mathfrak{a}\psi \subseteq v^{-1}\mathfrak{o}$; this shows $\mathfrak{a}\psi$ to be a fractional ideal.

Now it is clear from the definition that for any fractional ideal $\mathfrak{A}$ we have $\mathfrak{A}\phi\psi \supseteq \mathfrak{A}$ and by Lemma 4.3 equality holds, so that $\mathfrak{A}\phi\psi = \mathfrak{A}$. Next, if $\mathfrak{a} = \Pi \mathfrak{p}^{\alpha_\mathfrak{p}}$ and $\mathfrak{a}\psi = \mathfrak{A}$, then by Lemma 4.2 we can for any fixed $\mathfrak{p} \in S$ find $x \in \mathfrak{A}$ such that $v_\mathfrak{p}(x) = \alpha_\mathfrak{p}$. It follows that $v_\mathfrak{p}(\mathfrak{A}) = \alpha_\mathfrak{p}$ and so $\mathfrak{A}\phi = \mathfrak{a}$, i.e. $\mathfrak{a}\psi\phi = \mathfrak{a}$. Thus ϕ has the inverse ψ and hence is an isomorphism; in particular, it follows that F is a group. We sum up our conclusion as follows:

Theorem 4.4
Let K be a field, S a set of places on K satisfying the strong approximation property and let $\mathfrak{o}$ be the ring of integers for S. Then the fractional ideals in K relative to form a free abelian group under ideal multiplication and inversion, with a basis of integral ideals. Moreover, this group is isomorphic to the divisor group of S.■

An integral domain whose fractional ideals form a group under ideal multiplication is called a *Dedekind domain*. This is not the usual definition, but we shall soon prove the equivalence of this condition with other possibly more familiar forms (Theorem 4.6). For the moment we note that Theorem 4.4 shows that the ring of integers for a set of places with the strong approximation property is a Dedekind domain. It is of interest that the converse also holds.

We recall that a ring $\mathfrak{o}$ is called *Noetherian* (after E. Noether) if every ideal in $\mathfrak{o}$ is finitely generated, or equivalently, the ideals in $\mathfrak{o}$ satisfy the maximum condition: every non-empty set of ideals in $\mathfrak{o}$ includes a maximal member.

Theorem 4.5
Let $\mathfrak{o}$ be an integral domain with field of fractions K. If the fractional ideals of $\mathfrak{o}$ form a group under ideal multiplication, then $\mathfrak{o}$ can be defined as the intersection of principal valuation rings for a family of valuations with the strong approximation property.

Here we did not have to assume that the group of fractional ideals is free abelian; this emerges as a consequence.

Proof
We firstly note that $\mathfrak{o}$ is Noetherian. For if $\mathfrak{A}$ is any ideal in $\mathfrak{o}$, suppose that $\mathfrak{A} \neq 0$; then $\mathfrak{A}$ is invertible, so $\mathfrak{A}\mathfrak{A}^{-1} = \mathfrak{o}$, hence $1 = \Sigma a_i b_i$ for $a_i \in \mathfrak{A}$, $b_i \in \mathfrak{A}^{-1}$. It follows that any $x \in \mathfrak{A}$ can be written in the form

$$x = \Sigma a_i(b_i x),$$

and $b_i x \in \mathfrak{o}$, so $\mathfrak{A}$ is generated by $a_1, \ldots, a_n$.

If the different maximal ideals of $\mathfrak{o}$ are $\mathfrak{p}_1, \mathfrak{p}_2, \ldots$, then we have

$$\prod \mathfrak{p}_i^{\alpha_i} \subseteq \mathfrak{o} \quad \text{for any } \alpha_i \geqslant 0,$$

with equality if and only if $\alpha_i = 0$ for all i. Thus the $\mathfrak{p}_i$ generate a free abelian group F_0 say. If some integral ideal is not in F_0, then because $\mathfrak{o}$ is Noetherian we can find an ideal $\mathfrak{A}$ which is maximal among ideals not in F_0. We have $\mathfrak{A} \subseteq \mathfrak{p}_i \subset \mathfrak{o}$ for some maximal ideal $\mathfrak{p}_i$, hence $\mathfrak{A} \subset \mathfrak{A}\mathfrak{p}_i^{-1} \subseteq \mathfrak{o}$. By the maximality of $\mathfrak{A}$ it follows that $\mathfrak{A}\mathfrak{p}_i^{-1} \in F_0$ and so $\mathfrak{A} = \mathfrak{A}\mathfrak{p}_i^{-1}\mathfrak{p}_i \in F_0$, which is a contradiction. Thus F_0 contains every integral ideal and hence every principal ideal, for if $u = ab^{-1}$, then $a\mathfrak{o}, b\mathfrak{o} \in F_0$, hence also $u\mathfrak{o} = (a\mathfrak{o})(b\mathfrak{o})^{-1}$. If $\mathfrak{A}$ is any fractional ideal, then there exists $u \in \mathfrak{o}$, $u \neq 0$ such that $u\mathfrak{A} = \mathfrak{B}$ is integral, and hence in F_0; therefore $\mathfrak{A} = \mathfrak{B}u^{-1} \in F_0$ and this shows that F_0 includes all fractional ideals.

We have thus found that the group of all fractional ideals is free abelian, with the set of all maximal ideals as free generating set. Given $a \in K^{\times}$, the principal ideal $a\mathfrak{o}$ is a fractional ideal and so we have a representation

$$a\mathfrak{o} = \prod \mathfrak{p}^{\alpha_{\mathfrak{p}}(a)}. \tag{10}$$

For each maximal ideal $\mathfrak{p}$ the function $\alpha_{\mathfrak{p}}(a)$ is a principal valuation on K, as is easily verified. Let S be the set of places defined by these valuations. We have just seen that **D**.1 holds and **D**.2 follows because in (10) almost all the $\alpha_{\mathfrak{p}}(a)$ vanish, while for any fractional ideal $\mathfrak{A}$ we have $u\mathfrak{o} \subseteq \mathfrak{A} \subseteq v\mathfrak{o}$, hence

$$\alpha_{\mathfrak{p}}(u) \geqslant v_{\mathfrak{p}}(\mathfrak{A}) \geqslant \alpha_{\mathfrak{p}}(v),$$

so almost all the $v_{\mathfrak{p}}(\mathfrak{A})$ vanish. To prove **D**.3, take distinct maximal ideals $\mathfrak{p}$, $\mathfrak{q}$. Then $\mathfrak{p} + \mathfrak{q} = \mathfrak{o}$, hence $(\mathfrak{p} + \mathfrak{q})^{2N} = \mathfrak{o}$ for any $N > 0$, and so

$$\mathfrak{o} = (\mathfrak{p} + \mathfrak{q})^{2N} \subseteq \mathfrak{p}^{2N} + \mathfrak{p}^{2N-1}\mathfrak{q} + \ldots + \mathfrak{q}^{2N} \subseteq \mathfrak{p}^N + \mathfrak{q}^N.$$

It follows that we can write $1 = a + b$, where $a \in \mathfrak{p}^N$, $b \in \mathfrak{q}^N$ and $a, b \in \mathfrak{o}$. Thus a satisfies $\alpha_{\mathfrak{p}}(a) \geqslant N$, $\alpha_{\mathfrak{q}}(1 - a) \geqslant N$, $a \in \mathfrak{o}$, and so **D**.3 holds.■

For the sake of completeness we show the equivalence of the above definition of Dedekind domain with the more usual form.

Theorem 4.6 (*E. Noether*)
An integral domain $\mathfrak{o}$ with field of fractions K is a Dedekind domain if and only if it satisfies the following three conditions:

(i) *$\mathfrak{o}$ is Noetherian;*

(ii) *$\mathfrak{o}$ is integrally closed in K;*

(iii) *every non-zero prime ideal in $\mathfrak{o}$ is maximal.*

Proof
Let $\mathfrak{o}$ be a Dedekind domain. We saw in the proof of Theorem 4.5 that (i) holds, and clearly $\Pi\mathfrak{p}_i^{\alpha_i}$ is prime if and only if $\alpha_i \geqslant 0$ and $\Sigma\alpha_i = 1$; since each $\mathfrak{p}_i$ is maximal, (iii) follows. Now the localization $\mathfrak{o}_{\mathfrak{p}_i}$ is the valuation ring corresponding to the valuation defined by $\mathfrak{p}_i$, so we have $\mathfrak{o} = \cap\, \mathfrak{o}_{\mathfrak{p}_i}$ and (ii) will follow if we show that any element of K integral over $\mathfrak{o}$ belongs to $\mathfrak{o}_{\mathfrak{p}}$ for all $\mathfrak{p}$.

Given a maximal ideal $\mathfrak{p}$, let π be a prime element for the corresponding valuation $v_{\mathfrak{p}}$, thus $v_{\mathfrak{p}}(\pi) = 1$. Take $b \in K$ integral over $\mathfrak{o}$, say

$$b^n + a_1 b^{n-1} + \ldots + a_n = 0, \quad \text{where } a_i \in \mathfrak{o}; \tag{11}$$

we have to show that $b \in \mathfrak{o}_\mathfrak{p}$. Writing $b = \pi^r u$, where $v_\mathfrak{p}(u) = 0$, we have to show that $r \geqslant 0$. By substituting in (11) we find

$$u^n + a_1 u^{n-1} \pi^{-r} + \ldots + a_n \pi^{-rn} = 0.$$

If $r < 0$, this equation shows that $v_\mathfrak{p}(u) = v_\mathfrak{p}(u^n)/n > 0$, which is a contradiction. Therefore $r \geqslant 0$ and $b = \pi^r u \in \mathfrak{o}_\mathfrak{p}$, hence $b \in \cap\, \mathfrak{o}_\mathfrak{p} = \mathfrak{o}$ and (ii) follows.

Conversely, let $\mathfrak{o}$ be an integral domain with field of fractions K, satisfying (i)–(iii); we have to show that the fractional ideals form a group. In fact it is enough to show that every integral ideal is invertible. For given any fractional ideal $\mathfrak{A}$, let $u \in K^\times$ be such that $u\mathfrak{A} \subseteq \mathfrak{o}$. By hypothesis $u\mathfrak{A}$ has an inverse $\mathfrak{A}'$, thus $u\mathfrak{A}\mathfrak{A}' = \mathfrak{o}$, and so $u\mathfrak{A}'$ is the required inverse of $\mathfrak{A}$.

It remains to show that every integral ideal is invertible. We shall prove this in three steps:

1. Any non-zero ideal in $\mathfrak{o}$ contains a product of prime ideals. For if not, then by the Noetherian property there is a 'maximal offender' $\mathfrak{A}$. This ideal is not prime or $\mathfrak{o}$, so there exist $b_1, b_2 \in \mathfrak{o}$ such that $b_i \notin \mathfrak{A}$, $b_1 b_2 \in \mathfrak{A}$. Hence $\mathfrak{A} + b_i\mathfrak{o} \supset \mathfrak{A}$ and by the maximality of $\mathfrak{A}$ we can find prime ideals $\mathfrak{p}_1, \ldots, \mathfrak{p}_r$ such that $\mathfrak{A} + b_1\mathfrak{o} \supseteq \mathfrak{p}_1, \ldots \mathfrak{p}_s$, $\mathfrak{A} + b_2\mathfrak{o} \supseteq \mathfrak{p}_{s+1} \ldots \mathfrak{p}_r$. It follows that

$$\mathfrak{p}_1 \ldots \mathfrak{p}_r \subseteq (\mathfrak{A} + b_1\mathfrak{o})(\mathfrak{A} + b_2\mathfrak{o}) \subseteq \mathfrak{A},$$

and this contradicts the choice of $\mathfrak{A}$.
2. Any maximal ideal is invertible. Let $\mathfrak{p}$ be maximal and pick $a \in \mathfrak{p}$, $a \neq 0$. By 1. $a\mathfrak{o}$ contains a product of prime ideals, say

$$\mathfrak{p} \supseteq a\mathfrak{o} \supseteq \mathfrak{p}_1 \ldots \mathfrak{p}_r,$$

where r may be taken to be minimal. Since $\mathfrak{p}$ is prime, it contains some $\mathfrak{p}_i$, say $\mathfrak{p} \supseteq \mathfrak{p}_1$. But $\mathfrak{p}_1$ is also maximal, by (iii), so $\mathfrak{p}_1 = \mathfrak{p}$. Now $\mathfrak{p}_2 \ldots \mathfrak{p}_r \not\subseteq a\mathfrak{o}$, so for some $b \in \mathfrak{o}$, $b \in \mathfrak{p}_2 \ldots \mathfrak{p}_r$, $b \notin a\mathfrak{o}$ but $\mathfrak{p}b \subseteq a\mathfrak{o}$. Hence $a^{-1}b\mathfrak{p} \subseteq \mathfrak{o}$, i.e. $a^{-1}b \in (\mathfrak{o} : \mathfrak{p})$, but $a^{-1}b \notin \mathfrak{o}$. This shows that $(\mathfrak{o} : \mathfrak{p}) \neq \mathfrak{o}$. Now $\mathfrak{p} \subseteq \mathfrak{p}(\mathfrak{o} : \mathfrak{p}) \subseteq \mathfrak{o}$ and $\mathfrak{p}$ is maximal. If $\mathfrak{p}(\mathfrak{o} : \mathfrak{p}) = \mathfrak{p}$, then $(\mathfrak{o} : \mathfrak{p})$ is integral over $\mathfrak{o}$, by Lemma 2.2, and so, by (ii), $(\mathfrak{o} : \mathfrak{p}) \subseteq \mathfrak{o}$, but this is not the case. Hence $\mathfrak{p}(\mathfrak{o} : \mathfrak{p}) = \mathfrak{o}$ and so $\mathfrak{p}$ is indeed invertible.
3. Every integral ideal is invertible. For if not, then we can pick a maximal offender $\mathfrak{A}$. Clearly $\mathfrak{A} \subset \mathfrak{o}$, hence $\mathfrak{A} \subseteq \mathfrak{p} \subset \mathfrak{o}$ for some maximal ideal $\mathfrak{p}$, and thus $\mathfrak{A} \subset \mathfrak{A}\mathfrak{p}^{-1} \subseteq \mathfrak{o}$. Since $\mathfrak{A}$ is finitely generated and $\mathfrak{p}^{-1} \supset \mathfrak{o}$, it follows (again by Lemma 2.2) that $\mathfrak{A} \subset \mathfrak{A}\mathfrak{p}^{-1}$. By the maximality of $\mathfrak{A}$, $\mathfrak{A}\mathfrak{p}^{-1}$ has an inverse $\mathfrak{C}$, say: $\mathfrak{C}\mathfrak{A}\mathfrak{p}^{-1} = \mathfrak{o}$, and so $\mathfrak{p}^{-1}\mathfrak{C}$ is the inverse of $\mathfrak{A}$.■

We note that any principal ideal domain is a Dedekind domain, for the integral ideals are principal, hence invertible. Thus **Z** and the polynomial ring

$k[x]$ over any field are examples of Dedekind domains. However, the property of being a Dedekind domain is preserved under algebraic extensions, in contrast to what happens for principal ideal domains. This will be proved in section 2.5 and it will show that any finite ring of algebraic integers (i.e. the ring of integers in a finite algebraic extension) is a Dedekind domain.

Exercises

1. Let $\mathfrak{o}$ be an integral domain. Show that for any fractional ideals $\mathfrak{A}_1, \mathfrak{A}_2$ the sum $\mathfrak{A}_1 + \mathfrak{A}_2$ is again a fractional ideal, which in the case of principal ideals corresponds to the highest common factor.
2. Show that every ideal in Dedekind domain can be generated by two elements. (*Hint*: Use Theorem 4.5 and Lemma 4.2)
3. Let $K = k(x, y)$ be the rational function field in two indeterminates x, y. Show that any irreducible polynomial p in x, y defines a valuation on K and that the set S of all these valuations satisfies **D**.1 and **D**.2 but not **D**.3. Verify that the set of S-integers, the polynomial ring $k[x, y]$, is not a Dedekind domain.
4. Let L be the field of all algebraic numbers and S the set of all valuations on L. Show that S satisfies **D**.2 and **D**.3 but not **D**.1. Verify that the set of all algebraic integers is not a Dedekind domain.
5. Let $K = \mathbf{Q}(x)$ and extend the p-adic valuation v_p on $\mathbf{Q}$ to $\mathbf{Q}[x]$ by the rule: $v_p(\Sigma a_i x^i) = \min\{v(a_i) + i\}$. Let S be the family of all valuations on $\mathbf{Q}(x)$ obtained in this way from the p-adic valuations on $\mathbf{Q}$, for all primes p. Verify that **D**.1 and **D**.3 hold for S, but not **D**.2; show also that the ring of S-integers is not a Dedekind domain.
6. Let A be a Dedekind domain with field of fractions K. Show that any valuation ring in K containing A has the form of a localization $A_{\mathfrak{p}}$ (cf. section 1.7), where $\mathfrak{p}$ is a prime ideal in A.

2.5 EXTENSIONS OF DEDEKIND DOMAINS

Let L/K be a field extension. If a valuation v on K corresponds to a prime divisor $\mathfrak{p}$ and v has an extension w to L, with corresponding prime divisor $\mathfrak{P}$, then we shall write $\mathfrak{P} \mid \mathfrak{p}$ and say that $\mathfrak{P}$ *divides* $\mathfrak{p}$. The reason for this notation is that in terms of the corresponding fractional ideals (in the Dedekind case) we have $\mathfrak{p} \subseteq \mathfrak{P}$. We note that for a finite extension there are only a finite number of $\mathfrak{P}$ dividing a given place $\mathfrak{p}$, by Theorem 3.3. Consider a concrete case: let $K = \mathbf{Q}$, $L = \mathbf{Q}(\sqrt{p})$, where p is a prime number. The p-adic valuation on $\mathbf{Q}$ has uniformizer p and its extension to L has uniformizer $\pi = \sqrt{p}$. Here $e = 2, f = 1$ and symbolically we can write $p = \mathfrak{P}^2$, $\mathfrak{P} = (\sqrt{p})$.

Both $\mathbf{Z}$ and $k[x]$ are principal ideal domains, but their extensions in general are no longer principal. By contrast, the property of being a Dedekind domain is preserved by algebraic extensions, and this accounts for its importance. We shall give a proof in terms of the strong approximation property; for simplicity we limit ourselves to the case of separable extensions, although the result holds for any finite algebraic extensions (cf. Exercise 1).

Theorem 5.1

Let A be a Dedekind domain with field of fractions K, let L be a finite separable algebraic extension of K and denote by B the integral closure of A in L. Then B is again a Dedekind domain.

Proof

By Theorem 4.5, A can be defined by a set S_K of places for K with the strong approximation property:

$$A = \bigcap_{\mathfrak{p} \in S_K} \mathfrak{o}_{\mathfrak{p}}. \tag{1}$$

Denote by S_L the set of all places on L dividing a member of S_K. We shall show that S_L again has the strong approximation property and

$$B = \bigcap_{\mathfrak{P} \in S_L} \mathfrak{o}_{\mathfrak{P}}. \tag{2}$$

By Theorem 4.4 this will show that B is a Dedekind domain. We begin by proving (2). Given $\alpha \in B$, we take a monic equation for α over A:

$$\alpha^n + a_1\alpha^{n-1} + \ldots + a_n = 0, \tag{3}$$

where $a_i \in A$. Let $\mathfrak{P} \in S_L$ and suppose that $\alpha \notin \mathfrak{o}_{\mathfrak{P}}$; then $\alpha^{-1} \in \mathfrak{o}_{\mathfrak{P}}$ and we have

$$\alpha = -(a_1 + a_2\alpha^{-1} + \ldots + a_n\alpha^{1-n}) \in \mathfrak{o}_{\mathfrak{P}},$$

which is a contradiction. Hence $\alpha \in \mathfrak{o}_{\mathfrak{P}}$ and this holds for all $\mathfrak{P} \in S_L$, therefore $B \subseteq \bigcap \mathfrak{o}_{\mathfrak{p}}$. Conversely, if $\alpha \in \bigcap \mathfrak{o}_{\mathfrak{P}}$, then $v_{\mathfrak{P}}(\alpha) \geqslant 0$ for all $\mathfrak{P} \in S_L$. We take equation (3) to be the minimal equation for α over K and claim that $a_i \in A$; this will show that $\alpha \in B$.

If α' is a conjugate of α (in a normal closure of L over K), then $v_{\mathfrak{P}}(\alpha') = v_{\mathfrak{P}}'(\alpha)$ for some $\mathfrak{P}' \in S_L$, as we saw in the proof of Theorem 3.4, and $v_{\mathfrak{P}}'(\alpha) \geqslant 0$, hence $v_{\mathfrak{P}}(\alpha') \geqslant 0$. Thus all conjugates of α are integral at each $\mathfrak{P} \in S_L$ and since each a_i is a polynomial in the conjugates, it follows that $v_{\mathfrak{P}}(a_i) \geqslant 0$ for all i and all $\mathfrak{p} \in S_K$. Hence the coefficients of (3) lie in A, so α is integral over A and lies in B. This establishes (2).

It remains to show that S_L satisfies **D**.1–**D**.3. **D**.1 is clear from Theorem 1.1. To prove **D**.2, let α satisfy (3); there exist $\mathfrak{p}_1, \ldots, \mathfrak{p}_r \in S_K$ such that a_i is integral at all $\mathfrak{p} \neq \mathfrak{p}_j$. By Theorem 3.3, each $\mathfrak{p}_j$ has only finitely many extensions to L and α is integral at all $\mathfrak{P} \nmid \mathfrak{p}_j (j = 1, \ldots, r)$, so it is integral at almost all members of S_L and **D**.2 holds.

To verify **D**.3 we have to show: given $\mathfrak{P}', \mathfrak{P}'' \in S_L$ and $N > 0$, there exists $\alpha \in L$ such that

$$v_{\mathfrak{P}'}(\alpha-1)>N, \qquad v_{\mathfrak{P}''}(\alpha)>N, \quad v_{\mathfrak{P}}(\alpha)\geqslant 0 \quad \text{for all } \mathfrak{P}\neq\mathfrak{P}',\mathfrak{P}''.$$

Let $u_1, \ldots, u_n$ be a basis of L/K and choose $\mathfrak{p}_1, \ldots, \mathfrak{p}_r \in S_K$ such that

1. $v_{\mathfrak{P}}(u_i)\geqslant 0$ for $\mathfrak{P}\nmid\mathfrak{p}_j$, for all i, j;
2. $\mathfrak{P}', \mathfrak{P}''$ each divide some $\mathfrak{p}_j$ (not necessarily the same one).

It is clear, from **D**.2 for S_L, that this is possible. Let us fix $N>0$ and by the approximation theorem (Theorem 1.2.3) take $\beta\in L$ such that $v_{\mathfrak{P}'}(\beta-1)>N$, $v_{\mathfrak{P}''}(\beta)>N$, $v_{\mathfrak{P}}(\beta)\geqslant 0$ for $\mathfrak{P}\mid\mathfrak{p}_j$ for some j and $\mathfrak{P}\neq\mathfrak{P}',\mathfrak{P}''$. Now write $\beta=\Sigma y_iu_i$, take $M>0$ to be fixed later and (by **D**.3 for S_K) choose $x_i\in K$ such that

$$v_{\mathfrak{p}_j}(x_i-y_i)>M, \qquad v_{\mathfrak{p}}(x_i)\geqslant 0 \quad \text{for } \mathfrak{p}\neq\mathfrak{p}_1,\ldots,\mathfrak{p}_r \quad (i=1,\ldots,n, j=1,\ldots r).$$

Putting $\alpha=\Sigma x_iu_i$, we have

$$v_{\mathfrak{P}}(\alpha-\beta)=v_{\mathfrak{P}}(\Sigma(x_i-y_i)u_i)\geqslant \min_i\{M+v_{\mathfrak{P}}(u_i)\}, \quad \mathfrak{P}\mid\mathfrak{p}_1\ldots\mathfrak{p}_r,$$

$$v_{\mathfrak{Q}}(\alpha-\beta)\geqslant\min_i\{v_{\mathfrak{Q}}(u_i)\}\geqslant 0, \quad \mathfrak{Q}\nmid\mathfrak{p}_1\ldots\mathfrak{p}_r.$$

We now choose M so that $M>N-v_{\mathfrak{P}}(u_i)$ for all $\mathfrak{P}\mid\mathfrak{p}_1\ldots\mathfrak{p}_r$, $i=1,\ldots,n$. Then

$$v_{\mathfrak{P}'}(\alpha-1)\geqslant\min\{v_{\mathfrak{P}'}(\alpha-\beta), v_{\mathfrak{P}'}(\beta-1)\}>N,$$

$$v_{\mathfrak{P}''}(\alpha)\geqslant\min\{v_{\mathfrak{P}''}(\alpha-\beta), v_{\mathfrak{P}''}(\beta)\}>N,$$

$$v_{\mathfrak{Q}}(\alpha)\geqslant 0 \quad \text{for } \mathfrak{Q}\neq\mathfrak{P}',\mathfrak{P}''.$$

Thus S_L satisfies **D**.1–**D**.3 and this shows by Theorem 4.4 that B is Dedekind.■

Since every principal ideal domain is Dedekind, we have the following consequence:

Corollary 5.2
Any separable integral extension of a principal ideal domain is Dedekind.■

This shows the ring of integers in any finite algebraic extension to be Dedekind, and likewise for any function field of one variable; our proof covers the case of characteristic 0, the only one needed here.

Given a Dedekind domain A with field of fractions K, we have a homomorphism θ of $K^{\times}$ into the divisor group D_K, given by

$$\theta:\alpha\mapsto\prod\mathfrak{p}^{v_{\mathfrak{p}}(\alpha)}. \tag{4}$$

The kernel of this map is the group U of units in A, because the map from the fractional ideals to the divisors is an isomorphism, by Theorem 4.4. The cokernel is the *divisor class group* C_K, i.e. the group of classes of divisors modulo principal divisors. Thus we have an exact sequence

$$1 \to U \to K^{\times} \to D_K \to C_K \to 1. \tag{5}$$

As we have seen in section 2.4, the divisor group D_K is free abelian and this presents no problem. To describe the divisors in K more precisely we need information on U and C_K. For example, if we take $A = \mathbf{Z}$, then $K = \mathbf{Q}$ and we have $U = \{\pm 1\}$, $C = 1$. In general $C = 1$ means that every divisor, or equivalently (by Theorem 4.4), every fractional ideal is principal, so we have a principal ideal domain. This proves:

Proposition 5.3

For any Dedekind domain A, the divisor class group is trivial if and only if A is a principal ideal domain.■

In algebraic geometry divisors correspond to hypersurfaces of codimension 1, hence for the function field of a curve these hypersurfaces reduce to points, and then everything can be described by divisors. In this respect number fields are like function fields of curves, i.e. function fields of transcendence degree 1. We shall soon specialize to algebraic number fields and function fields of one variable, and two important results will describe the structure of C and U; in the former case:

1. C is finite;
2. U is a finitely generated abelian group. The elementary theory of such groups, together with the fact that every finite subgroup of a field is cyclic, shows that U has the form $U = \langle\zeta\rangle \times \mathbf{Z}^r$, where $\langle\zeta\rangle$ is cyclic of finite order and $\mathbf{Z}^r$ is free abelian of rank r.

In the function field case the situation is a little more complicated and we shall return to it in Chapter 4.

Consider again the map θ defined in (4). If L is a finite separable extension of K and θ' is the corresponding map for L, then for $\mathfrak{P}|\mathfrak{p}$ we have

$$v_{\mathfrak{P}}(a) = e_{\mathfrak{P}} v_{\mathfrak{p}}(a) \quad \text{for } a \in K, \tag{6}$$

where $e_{\mathfrak{P}}$ is the ramification index for $\mathfrak{P}$. Hence for any $a \in K$ we have

$$a\theta' = \prod_{\mathfrak{P}} \mathfrak{P}^{v_{\mathfrak{P}}(a)} = \prod_{\mathfrak{p}} \left(\prod_{\mathfrak{P}|\mathfrak{p}} \mathfrak{P}^{v_{\mathfrak{P}}(a)} \right) = \prod_{\mathfrak{p}} \left(\prod_{\mathfrak{P}|\mathfrak{p}} \mathfrak{P}^{e_{\mathfrak{P}}} \right)^{v_{\mathfrak{p}}(a)}.$$

Thus we obtain a commutative diagram as shown below by defining a map $D_K \to D_L$

$$\mathfrak{p} \mapsto \prod_{\mathfrak{P}|\mathfrak{p}} \mathfrak{P}^{e_{\mathfrak{P}}}. \tag{7}$$

This is an embedding of D_K in D_L, sometimes called the *conorm map*.

In the other direction we have the *norm map*. Given a place $\mathfrak{p}$ of K, let the

$$\begin{array}{ccc} L & \xrightarrow{\theta'} & D_L \\ \uparrow & & \uparrow \\ K & \xrightarrow[\theta]{} & D_K \end{array}$$

divisors of $\mathfrak{p}$ in L be $\mathfrak{P}_1, \ldots, \mathfrak{P}_r$ and let $L_1, \ldots, L_r$ be the corresponding extensions of $\hat{K}$, the completion of K (cf. Theorem 3.3). As we saw, we have $[L_i : \hat{K}] = n_i = e_i f_i$, hence

$$v_{\mathfrak{P}_i}(\alpha) = \frac{1}{n_i} v_{\mathfrak{P}_i}(N_{L_i/\hat{K}}(\alpha)) = (e_i/n_i) v_{\mathfrak{p}}(N_{L_i/\hat{K}}(\alpha)),$$

and so we find that

$$v_{\mathfrak{p}}(N_{L/K}(\alpha)) = \sum v_{\mathfrak{p}}(N_{L_i/\hat{K}}(\alpha)) = \sum f_i v_{\mathfrak{P}_i}(\alpha).$$

In terms of divisors this may be written as

$$N_{L/K}(\mathfrak{P}_i) = \mathfrak{p}^{f_i}. \tag{8}$$

Hence we obtain

$$N(\mathfrak{p}) = N(\mathfrak{P}_1^{e_1} \ldots \mathfrak{P}_r^{e_r}) = \mathfrak{p}^{\Sigma e_i f_i} = \mathfrak{p}^n,$$

where we have used Theorem 3.3. Thus we see that norm(conorm) $= n$, while of course conorm (norm) $= 1$.

To conclude this section we record another criterion for a Dedekind domain to be principal which is often useful. We recall that in an integral domain an *atom* or *irreducible element* is a non-unit which cannot be written as a product of two non-units; a *prime* is an element p, neither zero nor a unit, such that $p|ab$ implies $p|a$ or $p|b$. It is easily seen that every prime is an atom, but the converse need not hold. In fact, the converse, together with a finiteness condition, is equivalent to unique factorization of elements. By a *unique factorization domain* (UFD) one understands an integral domain in which every element not zero or a unit can be written as a product of atoms:

$$c = a_1 \ldots a_r, \tag{9}$$

and if $c = b_1 \ldots b_s$ is another such factorization, then $s = r$ and the b's can be renumbered so that b_i is associated to a_i. It can be shown that an integral domain is a UFD if and only if every element c not zero or a unit has a factorization (9) into atoms and every atom is prime (cf. A. 1, 6.5).

Proposition 5.4
A Dedekind domain is principal if and only if it is a unique factorization domain.

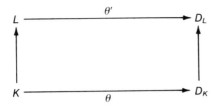

Proof.
It is well known that every principal ideal domain has unique factorization (cf. e.g. A.1, Theorem 3 of 10.5). Conversely, assume that A is a UFD as well as Dedekind, and take any prime ideal $\mathfrak{p} \neq 0$. If $c = ab \in \mathfrak{p}$, then $a \in \mathfrak{p}$ or $b \in \mathfrak{p}$, because $\mathfrak{p}$ is a prime ideal. Hence $\mathfrak{p}$ contains a prime element c. Now (c) is a non-zero prime ideal contained in $\mathfrak{p}$, hence $(c) = \mathfrak{p}$ by Theorem 4.6. Thus every prime ideal is principal; since every ideal is a product of prime ideals, every ideal is principal, as we had to show.■

Exercises

1. Let K be a field of prime characteristic p, with an element a and form $L = K(a^{1/q})$, where $q = p^r$. Given a Dedekind domain A in K with K as field of fractions, show that its integral closure in L is again Dedekind. Deduce that Theorem 5.1 holds without the separability assumption.
2. Let L be a field with a group G of automorphisms and let B be a Dedekind domain with L as field of fractions, and admitting the automorphisms of G. If K is the fixed field of G and $A = B \cap K$, show that A is a Dedekind domain with field of fractions K.
3. Let $\mathfrak{o}$ be the ring of integers in an algebraic number field. Show that for any integral ideal $\mathfrak{a}$, the residue class ring $\mathfrak{o}/\mathfrak{a}$ is finite, with $N(\mathfrak{a})$ elements. (*Hint:* Factorize $\mathfrak{a}$ and use (8).)

2.6 DIFFERENT AND DISCRIMINANT

In solving polynomial equations many of the problems arise from the presence of repeated factors (the inseparable case); this situation can usually be recognized by examining the derivative. Similarly, in the study of extensions of Dedekind domains there is a critical ideal which indicates which places are ramified. This is the discriminant; for a clearer understanding we first define the different, from which the discriminant can be formed by taking norms.

Let L/K be a finite algebraic extension of fields. We recall that there is a notion of trace defined on L, $\mathrm{Tr}(x)$, as the sum of all the conjugates of x (in an algebraic closure of L), with values in K, such that

$$\mathrm{Tr}(x+y) = \mathrm{Tr}(x) + \mathrm{Tr}(y), \qquad \mathrm{Tr}(\lambda x) = \lambda \mathrm{Tr}(x),$$
$$\mathrm{Tr}(1) = [L:K] \quad (x, y \in L, \lambda \in K).$$

Moreover, the bilinear form $\mathrm{Tr}(xy)$ defined by the trace is non-singular whenever L/K is separable (cf. Proposition 3.9.5 of A.2); this means that $\mathrm{Tr}(ax) = 0$ for all $x \in L$ implies $a = 0$.

We shall assume L/K to be separable, and take A to be a Dedekind domain with K as its field of fractions and B its integral closure in L. By Corollary 5.2, B is again a Dedekind domain.

Given $X \subseteq L$, we define its *complementary set* or simply *complement* as

$$X' = \{y \in L \mid \mathrm{Tr}(yX) \subseteq A\}. \tag{1}$$

Clearly X' is a subgroup of the additive group of L. Moreover, we have $\mathrm{Tr}(yz\,.\,X) = \mathrm{Tr}(y\,.\,zX)$, hence

$$(zX)' = z^{-1}X', \tag{2}$$

$$X \subseteq Y \Rightarrow X' \supseteq Y', \tag{3}$$

$$zX \subseteq X \Rightarrow zX' \subseteq X' \tag{4}$$

Here (2) and (3) are immediate, and (4) follows by combining them. Let $u_1, \ldots, u_n$ be a basis of L/K. Since Tr is non-singular, the mapping

$$x \mapsto (\mathrm{Tr}(u_1x), \ldots, \mathrm{Tr}(u_nx)),$$

as mapping of n-dimensional spaces, is injective and hence an isomorphism. Thus we can find $v_1, \ldots, v_n \in L$ to satisfy

$$\mathrm{Tr}(u_iv_j) = \delta_{ij}. \tag{5}$$

The v's form the *dual basis* for the u's. With its help we can easily describe the complement of an n-dimensional free A-submodule of L:

Lemma 6.1
Let K, L, A, B be as above and $[L : K] = n$. If U is a free A-submodule of L of rank n, then its complement U' is also free of rank n, and we obtain a basis of U' by taking the dual of a basis of U.

Proof
Let $u_1, \ldots, u_n$ be an A-basis of U. Then $u_1, \ldots, u_n$ is also a basis for L over K, and as we have seen, there is a dual basis $v_1, \ldots, v_n$ satisfying (5). We claim that $v_1, \ldots, v_n$ is an A-basis for U'.

Any $x \in U$ has the form $x = \Sigma a_iu_i$, where $a_i \in A$ and by (5), $\mathrm{Tr}(xv_j) = a_j \in A$; it follows that $v_1, \ldots, v_n \in U'$. Conversely, given $y \in U'$, we can write $y = \Sigma\beta_iv_i$, where $\beta_i \in K$, and by definition of y, $\beta_i = \mathrm{Tr}(u_iy) \in A$, hence y lies in the A-module spanned by the v's, so the v's form an A-basis for U'.■

In general a fractional ideal in L need not be free as A-module; this need not even hold for B itself, but B is free whenever A is a principal ideal domain. For the proof we shall need the well-known result that any submodule of a free A-module of rank n is again free of rank at most n (cf. A.1, Proposition 1 of 10.6).

Theorem 6.2
Let A be a principal ideal domain with field of fractions K, L a finite separable extension of K and B the integral closure of A in L. Then B and generally every fractional ideal of B is a free A-module of rank $[L : K]$.

Proof

Let $u_1, \ldots, u_n$ be a basis of L over K, $n = [L:K]$. On multiplying the u's by suitable elements we may assume that $u_i \in B$ $(i = 1, \ldots, n)$. Then the A-module U spanned by the u's satisfies $U \subseteq B$, hence $U' \supseteq B'$ by (3). Since $\mathrm{Tr}(B \,.\, B) \subseteq A$, we have $B \subseteq B'$ and so

$$U \subseteq B \subseteq B' \subseteq U'.$$

By Lemma 6.1, U' is free of rank n, hence B as submodule of U' is also free of rank at most n. In fact the rank must be n because B contains U of rank n. Now the result follows for any fractional ideal $\mathfrak{A}$ because we have $cB \subseteq \mathfrak{A} \subseteq dB$, hence

$$d^{-1}B' \subseteq \mathfrak{A}' \subseteq c^{-1}B'. \blacksquare$$

We also note the explicit formula for the complement of any invertible ideal $\mathfrak{A}$ (not necessarily principal):

$$\mathfrak{A}' = B'\mathfrak{A}^{-1}. \tag{6}$$

For we have $\mathrm{Tr}(B'\mathfrak{A}^{-1}\mathfrak{A}) = \mathrm{Tr}(B'B) \subseteq A$, hence $B'\mathfrak{A}^{-1} \subseteq \mathfrak{A}'$, and conversely,

$$\mathrm{Tr}(\mathfrak{A}'\mathfrak{A}B) \subseteq \mathrm{Tr}(\mathfrak{A}'\mathfrak{A}) \subseteq A,$$

hence $\mathfrak{A}'\mathfrak{A} \subseteq B'$ and so $\mathfrak{A}' \subseteq B'\mathfrak{A}^{-1}$, which proves equality in (6). Here it is not necessary for the ground ring to be principal, but merely for the extension L/K to be separable. In that case we can express the relation (6) as follows.

Proposition 6.3

Let L/K be a separable extension, A a Dedekind ring with field of fractions K and B its integral closure in L. Then its complement B' is an invertible ideal whose complement is integral:

$$\mathfrak{d} = (B')^{-1}, \tag{7}$$

and for any fractional ideal $\mathfrak{A}$ of B, its inverse is related to its complement by the formula

$$\mathfrak{A}^{-1} = \mathfrak{d}\mathfrak{A}'. \tag{8}$$

Proof

We have seen that $B' \supseteq B$ and it is clear that $BB' \subseteq B'$, by (4). Moreover, B' is finitely generated as A-module, by $c_1, \ldots, c_r$ say. Writing the c_i as fractions with a common denominator in B, say $c_i = a_i b^{-1}$, where $a_i, b \in B$, we have $B' \subseteq b^{-1}B$ and this shows B' to be a fractional ideal. Since $B' \supseteq B$, we have $(B')^{-1} \subseteq B$ and so $\mathfrak{d}$ defined by (7) is an integral ideal. Now (8) follows on multiplying both sides of (6) by $\mathfrak{d}$. $\blacksquare$

The ideal $\mathfrak{d}$ defined by (7) is called the *different* of the extension B over A

(or also of L over K). In the particular case where B is generated by a single element over A, there is an explicit formula for $\mathfrak{d}$ which also explains the name.

Theorem 6.4
Let A be a Dedekind domain, $B = A[\alpha]$ a separable extension generated by a single element α integral over A, and let f be the minimal polynomial for α over A. Then $B' = B/(f'(\alpha))$, where f' denotes the derivative, and so

$$\mathfrak{d} = (f'(\alpha)). \tag{9}$$

Proof
Let $\alpha_1, \ldots, \alpha_n$ be the roots of $f = 0$ in an algebraic closure of A; by separability they are distinct. The Lagrange interpolation formula shows that we have

$$X^r = \sum_i \frac{f(X)}{X - \alpha_i} \cdot \frac{\alpha_i^r}{f'(\alpha_i)} \quad (r = 0, 1, \ldots, n-1). \tag{10}$$

This is easily verified by observing that both sides are polynomials in X of degree less than n which agree at the n points $\alpha_1, \ldots, \alpha_n$. In terms of the trace we can rewrite it as

$$X^r = \mathrm{Tr}\left(\frac{f(X)}{X - \alpha} \cdot \frac{\alpha^r}{f'(\alpha)}\right),$$

where Tr is applied to the coefficients only. If we write

$$f(X)/(X - \alpha) = b_0 + b_1 X + \ldots + b_{n-1} X^{n-1}, \tag{11}$$

where $b_i \in L$, then by equating coefficients in equation (10), we find

$$\mathrm{Tr}(b_j \alpha^r / f'(\alpha)) = \delta_{jr}.$$

It follows that the elements $b_j/f'(\alpha)$ form a dual basis for $1, \alpha, \ldots, \alpha^{n-1}$. To complete the proof we observe that if

$$f = X^n + a_1 X^{n-1} + \ldots + a_n,$$

then

$$f(X)/(X - \alpha) = X^{n-1} + (a_1 + \alpha) X^{n-2} + (a_2 + a_1\alpha + \alpha^2) X^{n-3} + \ldots$$
$$+ (a_{n-1} + a_{n-2}\alpha + \ldots + \alpha^{n-1})$$

A comparison with equation (11) shows that each of the sets $b_o, \ldots, b_{n-1}$ and $1, \alpha, \ldots, \alpha^{n-1}$ can be expressed as a linear combination of the other with coefficients from A. Hence the b's form another A-basis for B, and since the $b_j/f'(\alpha)$ form a basis for B', we find that $B = (f'(\alpha))B'$, from which equation (9) follows.■

We now turn to discriminants; it will be convenient to define them for general subspaces at first. Let L/K again be a separable extension with rings of integers B, A, and consider any A-submodule U of L which is free of rank $n=[L:K]$ as A-module. We define the *discriminant* $D(U)$ of U relative to any A-basis $u_1, \ldots, u_n$ of U as

$$D(U)=\det(\mathrm{Tr}(u_iu_j)). \tag{12}$$

If $u_1', \ldots, u_n'$ is another basis of U, related to the first by equations

$$u_i'=\Sigma\, p_{ij}u_j,$$

then $\mathrm{Tr}(u_i'u_j')=\Sigma p_{ir}p_{js}\mathrm{Tr}(u_ru_s)$, hence the discriminant $D'(U)$ relative to $u_1', \ldots, u_n'$ is related to $D(U)$ by the equation

$$D'(U)=\det(p_{ij})^2D(U). \tag{13}$$

The matrix $P=(p_{ij})$ has entries in A and is invertible, hence $\det P$ is a unit in A and so D' differs from D by the square of a unit. In particular, if B is a free A-module, this defines the *discriminant* of B, given by (12) with $U=B$, up to a square of a unit in terms of a basis of B. For a separable extension it is always non-zero; it can be shown that the discriminant (suitably defined) of an inseparable extension vanishes.

To facilitate the study of the discriminant we shall now obtain another expression for it. Since L/K was assumed separable, there are $n=[L:K]$ homomorphisms $\sigma_1, \ldots, \sigma_n$ of L over K into an algebraic closure of L. For any $x\in L$ we have $\mathrm{Tr}(x)=\Sigma x^{\sigma_i}$, hence

$$D(U)=\det(\mathrm{Tr}(u_iu_j))=\det PP^T, \tag{14}$$

where $P=(p_{ij})$, $p_{ij}=u_i^{\sigma_j}$, and P^T indicates the transpose matrix.

If we replace $u_1, \ldots, u_n$ by $u_1\alpha, \ldots, u_n\alpha$, where $\alpha\in L$, then P becomes PC, where $C=\mathrm{diag}\,(\alpha^{\sigma_1}, \ldots, \alpha^{\sigma_n})$. It is clear that $\det C$ is the product of the conjugates of α, i.e. $N(\alpha)$. Hence we have, on writing $U\alpha$ for the space with basis $u_1\alpha, \ldots, u_n\alpha$,

$$D(U\alpha)=\det(PCC^TP^T)=N(\alpha)^2D(U). \tag{15}$$

Now we can introduce the norm of an ideal as follows. Let $\mathfrak{A}=\prod\mathfrak{P}_i^{\alpha_i}$ be any fractional ideal in B. For a given place $\mathfrak{P}_i$ we defined its norm in (8) of section 2.5 as

$$N(\mathfrak{P}_i)=\mathfrak{p}^{f_i}, \tag{16}$$

where $\mathfrak{p}$ is the place of K which $\mathfrak{P}_i$ divides and f_i is the corresponding residue degree. Thus we may define the norm of $\mathfrak{A}$ by

$$N(\mathfrak{A})=\prod N(\mathfrak{P}_i)^{\alpha_i}. \tag{17}$$

Here (16) is the 'local' definition, which suffices to give the general case (17).

We apply this result to (15), taking $U=B$, $U\alpha=B'$. Locally, i.e. at a place

$\mathfrak{P}_i$, we can do this because in the corresponding valuation ring every ideal is principal; we thus obtain

$$D(B') = N(B')^2 D(B). \tag{18}$$

Suppose now that $\{u_i\}$, $\{v_j\}$ are dual bases for B, B' respectively and put $p_{ij} = u_i^{\sigma_j}$, $q_{ij} = v_i^{\sigma_j}$, $P = (p_{ij})$, $Q = (q_{ij})$. Then

$$\sum_r p_{ir} q_{jr} = \mathrm{Tr}(u_i v_j) = \delta_{ij}$$

hence $PQ = I$, and since $D(B) = (\det P)^2$, $D(B') = (\det Q)^2$ by (14), we have $D(B)D(B') = 1$. Combining this with (18), we find

$$D(B')^2 = N(B')^2.$$

Now the discriminant $\mathfrak{D}$ of L/K is $D(B)$, hence we have

$$\mathfrak{D} = D(B')^{-1} = N(B')^{-1}.$$

Thus in terms of fractional ideals we have the equation:

$$\mathfrak{D} = N(\mathfrak{d}), \tag{19}$$

which identifies the discriminant as the norm of the different.

To avoid ambiguity in dealing with several extensions we shall also write $\mathfrak{D}_{L/K}$ for $\mathfrak{D}$ and $\mathfrak{d}_{L/K}$ for $\mathfrak{d}$. For repeated extensions we have the following formulae:

Proposition 6.5

Let A be a Dedekind domain with field of fractions K, let $K \subseteq L \subseteq M$ be finite separable algebraic extensions and B, C the integral closures of A in L, M respectively. Then

$$\mathfrak{d}_{M/K} = \mathfrak{d}_{M/L}\mathfrak{d}_{L/K}, \tag{20}$$

$$\mathfrak{D}_{M/K} = N_{L/K}(\mathfrak{D}_{M/L}) \cdot \mathfrak{D}_{L/K}^{[M:L]}. \tag{21}$$

Proof

In order to establish equation (20) we must show that

$$C'_{M/K} = C'_{M/L} B'_{L/K}, \tag{22}$$

where the prime means the complement in the extension indicated by the suffix. We have $\mathrm{Tr}_{M/K}(x) = \mathrm{Tr}_{L/K}(\mathrm{Tr}_{M/L}(x))$, hence

$$\mathrm{Tr}_{M/K}(C'_{M/L}B'_{L/K}C) = \mathrm{Tr}_{L/K}(B'_{L/K}\mathrm{Tr}_{M/L}(C'_{M/L}C))$$

$$\subseteq \mathrm{Tr}_{L/K}(B'_{L/K}B) \subseteq A.$$

This shows that the left-hand side of (22) includes the right-hand side. To obtain the reverse inclusion, take $\gamma \in C'_{M/K}$; then

$$\mathrm{Tr}_{L/K}(B\,.\,\mathrm{Tr}_{M/L}(\gamma C)) \subseteq \mathrm{Tr}_{M/K}(\gamma C) \subseteq A,$$

hence $\mathrm{Tr}_{M/L}(\gamma C) \subseteq B'_{L/K}$ and so

$$\mathrm{Tr}_{M/L}(B'^{-1}_{L/K} \cdot \gamma C) = B'^{-1}_{L/K} \cdot \mathrm{Tr}_{M/L}(\gamma C) \subseteq B.$$

Therefore $B'^{-1}_{L/K}\gamma \subseteq C'_{M/L}$ and so $\gamma \in C'_{M/L}\, B'_{L/K}$, which shows that equality holds in (22). This proves (20), and now (21) follows by taking norms.■

We remark that the different and discriminant can also be defined locally. Omitting the field suffix, we have

$$\mathfrak{d} = \Pi \mathfrak{d}_{\mathfrak{P}}, \qquad \mathfrak{D} = \Pi \mathfrak{D}_{\mathfrak{p}}.$$

Here $\mathfrak{d}_{\mathfrak{P}} = \mathfrak{d}_{L_{\mathfrak{P}}/K_{\mathfrak{p}}}$, is the local different, in contrast to the global version $\mathfrak{d} = \mathfrak{d}_{L/K}$, and similarly for the discriminant.

We illustrate the results by looking at some quadratic extensions:

(i) $\mathbf{Q}(\sqrt{d})$. There are two cases:

(a) $d \equiv 2$ or 3 (mod 4). Basis 1, $\sqrt{d}$

$$D = \begin{vmatrix} 1 & \sqrt{d} \\ 1 & -\sqrt{d} \end{vmatrix}^2 = 4d.$$

(b) $d \equiv 1 (\mathrm{mod}\, 4)$. Basis $1, u = (1 + \sqrt{d})/2$. We also put $u* = (1 - \sqrt{d})/2$.

$$D = \begin{vmatrix} 1 & u \\ 1 & u^* \end{vmatrix}^2 = d.$$

(ii) Let k be a field of characteristic 2 and a an element of k, not a square. The extension $k(\sqrt{a})$ has the basis 1, α, where $\alpha^2 = a$. Here Tr vanishes identically, so the discriminant, defined as det $(\mathrm{Tr}(u_i u_j))$ for a basis $\{u_i\}$, is zero. Likewise the different is zero since the minimal polynomial for α has derivative zero.

(iii) $K = \mathbf{C}(x)$, $L = \mathbf{C}(x, y)$, where $x^2 + y^2 = 1$. A basis for L/K is 1, y and

$$D = \begin{vmatrix} 1 & y \\ 1 & -y \end{vmatrix}^2 = 4(1 - x^2).$$

In each case the ramification (measuring the increase in the value group) is measured by the divisors of $\mathfrak{D}$. We also note that in example (iii) the ramification can be avoided by choosing a suitable uniformizer such as $t = y/(1 + x)$, in terms of which

$$x = \frac{1 - t^2}{1 + t^2}, \qquad y = \frac{2t}{1 + t^2}.$$

Here t is a 'global' uniformizer, which exists because $\mathbf{C}(x, y)$ is purely transcendental, in contrast to the 'local' uniformizers, which always exist for a principal valuation (cf. section 1.4).

To conclude this section we prove the *Dedekind discriminant theorem*, which provides more precise information about the discriminant divisors.

Theorem 6.6

Let L/K be a finite separable algebraic extension, A a Dedekind domain with field of fractions K and B the integral closure of A in L, and express the discriminant as a product of prime divisors in A:

$$\mathfrak{D}_{L/K} = \prod \mathfrak{p}^{\delta_\mathfrak{p}}.$$

For any prime divisor $\mathfrak{p}$ of A let $\mathfrak{P}_1, \ldots, \mathfrak{P}_r$ be its extensions to B and write e_i, f_i for the ramification index and residue degree of the $\mathfrak{P}_i$-adic valuation. Then

$$\delta_\mathfrak{p} \geqslant \Sigma f_i(e_i - 1). \tag{23}$$

Proof

It is enough to prove this locally, so we may assume that $A = \mathfrak{o}_\mathfrak{p}$, $B = B_1 \cap \ldots \cap B_r$, where $B_i = \mathfrak{o}_{\mathfrak{P}_i}$ for short. Let p, π_i be uniformizers for $\mathfrak{p}$, $\mathfrak{P}_i$ respectively, so that

$$p = \varepsilon \pi_1^{e_1} \ldots \pi_r^{e_r},$$

where ε is a unit in B. Now take elements $u_{i\nu}(\nu = 1, \ldots, f_i)$ in B_i such that their residues mod $\mathfrak{P}_i$ form a basis of $B_i/\mathfrak{P}_i$ over $A/\mathfrak{p}$. Secondly fix a positive integer h and take elements

$$w_{i\mu} = \pi_i^\mu (\pi_1^{e_1} \ldots \hat{\pi}_i^{e_i} \ldots \pi_r^{e_r})^h, \quad \mu = 0, 1, \ldots, e_i - 1,$$

where ^ indicates a term to be omitted. By choosing h large enough, we can ensure that all the elements

$$w_{i\mu} u_{i\nu}, \quad (i = 1, \ldots, r, \mu = 0, \ldots, e_i - 1, \nu = 1, \ldots, f_i) \tag{24}$$

lie in each B_j and hence in B. We claim that (24) is a basis for B over A. In the first place, if $\alpha \in B$, then for each i there exist elements $a_{io\nu} \in A$ such that

$$\alpha \equiv \sum_\nu a_{io\nu} w_{io} u_{i\nu} \pmod{\mathfrak{P}_i}. \tag{25}$$

Hence $(\alpha - \Sigma a_{io\nu} w_{io} u_{i\nu}) w_{i1}^{-1} \in B_i$ and so there exist $a_{i1\nu} \in A$ such that

$$(\alpha - \Sigma a_{io\nu} w_{io} u_{i\nu}) w_{i1}^{-1} \equiv \Sigma a_{i1\nu} w_{i1} u_{i\nu} \pmod{\mathfrak{P}_i}.$$

Therefore

$$\alpha \equiv \sum_{\mu=0}^{1} a_{i\mu\nu} w_{i\mu} u_{i\nu} \pmod{\mathfrak{P}_i^2}.$$

By induction we obtain after e_i steps

$$\alpha \equiv \sum_{\mu\nu} a_{i\mu\nu} w_{i\mu} u_{i\nu} \pmod{\mathfrak{P}_i^{e_i}}.$$

If we choose h so large that $h > e_j$ for all j, then

$$\sum_{\mu\nu} a_{i\mu\nu} w_{i\mu} u_{i\nu} \equiv 0 \pmod{\mathfrak{P}_j^{e_j}} \quad \text{for } j \neq i.$$

Writing $\alpha_i = \Sigma a_{i\mu\nu} w_{i\mu} u_{i\nu}$, we thus have $\alpha - \alpha_i \in \mathfrak{P}_i^{e_i}$, $\alpha_i \in \mathfrak{P}_j^{e_j}$ for $j \neq i$. Hence

$$\alpha - \sum \alpha_i \in \prod \mathfrak{P}_i^{e_i} = \mathfrak{p},$$

and so

$$\alpha \equiv \sum_{i\mu\nu} a_{i\mu\nu} w_{i\mu} u_{i\nu} \pmod{B\mathfrak{p}}. \tag{26}$$

Let V be the A-module spanned by all the $w_{i\mu} u_{i\nu}$; by (26) we have $B = V + B\mathfrak{p}$, but $B\mathfrak{p}$ is contained in the Jacobson radical of B, and so by Nakayama's lemma (cf. A.2, Lemma 5.4.6), $V = B$, so that the elements (24) span B as A-module.

To verify this directly and to show that the elements (24) form a basis (over A) we note that the $a_{i\mu\nu}$ in (26) are unique mod $\mathfrak{p}$, from which it follows that the elements (24) are K-linearly independent (mod $\mathfrak{p}$) and hence K-linearly independent. Their number is $\Sigma e_i f_i = [L : K]$ (by Theorem 3.3), so they form a K-basis for L. Now A is a principal valuation ring, and so B has a basis over A, say $\{v_\lambda\}$, which can of course be expressed as a K-linear combination of the $w_{i\mu} u_{i\nu}$. With a common denominator c we can write

$$v_\lambda = c^{-1} \cdot \sum c_{\lambda, i\mu\nu} w_{i\mu} u_{i\nu}, \quad \text{where } c, c_{\lambda, i\mu\nu} \in A. \tag{27}$$

We may assume that not all the $c_{\lambda, i\mu\nu}$ are divisible by $\mathfrak{p}$, but that c is. Multiplying up, we obtain

$$\sum c_{\lambda, i\mu\nu} w_{i\mu} u_{i\nu} \equiv 0 \pmod{\mathrm{B}\mathfrak{p}},$$

which is a contradiction. Hence $c = 1$ in (27) and it follows that the $w_{i\mu} u_{i\nu}$ form an A-basis for B.

Now the discriminant can be expressed as

$$\mathfrak{D} = \det(\mathrm{Tr}(w_{i\mu} u_{i\nu} w_{j\rho} u_{j\sigma})).$$

If $i \neq j$, then $\mathrm{Tr}(w_{i\mu}u_{i\nu}w_{j\rho}u_{j\sigma}) \equiv 0 \pmod{\mathfrak{p}^h}$, and h can be increased, if necessary. Thus we may assume that $h > \delta_\mathfrak{p}$ and take the determinant mod $\mathfrak{p}^h$; then it reduces to block form and so becomes a product of determinants with fixed i, and it is enough to consider one of these, say $i = 1$. Henceforth we suppress the corresponding suffix. We thus have to consider

$$\det(\mathrm{Tr}(w_\mu u_\nu w_\rho u_\sigma)). \tag{28}$$

This is the determinant of a matrix with $e_1 f_1$ rows and columns.
Now for any $\alpha \in B$ and any $\mu = 0, 1, \ldots, e_1 - 1$ we have

$$\mathrm{Tr}(\alpha w_\mu) \equiv 0 \pmod{\mathfrak{p}}. \tag{29}$$

For $(\alpha w_\mu)^{e_1} \equiv 0 \pmod{\mathfrak{p}}$, so αw_μ is nilpotent mod $\mathfrak{p}$ and hence its trace is 0. The determinant (28) therefore has the following form, where we have taken $e_1 = 3$ for purposes of illustration:

$$\begin{matrix} f_1 \\ f_1 \\ f_1 \end{matrix} \begin{pmatrix} * & 0 & 0 \\ 0 & 0 & 0 \\ 0 & 0 & 0 \end{pmatrix} \pmod{\mathfrak{p}}.$$

All rows after the first f_1 are $\equiv 0 \pmod{\mathfrak{p}}$, and there are $(e_1 - 1)f_1$ in all. Each contributes a power of $\mathfrak{p}$ to $\mathfrak{D}$, hence $\delta_\mathfrak{p} \geqslant \Sigma(e_i - 1)f_i$ and the result follows.■

This result shows that the number of places of A that can be ramified in B is bounded by the exponent of the discriminant. Since only finitely many of these exponents are positive, the number of ramified places must also be finite:

Corollary 6.7
For A, B as in Theorem 6.6, $\mathfrak{P}|\mathfrak{p}$ is unramified for almost all places $\mathfrak{p}$ of A.■

For Galois extensions of $\mathbf{Q}$ the inequality of Theorem 6.6 can be replaced by a precise formula for δ_p, using Hilbert's theory of ramification. There is then also a similar result for the different: locally we have $\mathfrak{d} = \mathfrak{P}^\nu$, where $\nu \geqslant e - 1$, with equality if and only if e is not divisible by the rational prime dividing $\mathfrak{P}$ (i.e. $\mathfrak{P}$ is 'tamely ramified'), cf. Hasse (1980), Chapter 25.

Exercises

1. For L/K separable and A, B as in the text, show that for any subspace U of L, free as A-module of rank $[L : K]$, $U'' = U$.
2. Let L/K be a finite separable extension. Show that the different of L/K can be written as $\Pi(\alpha_i - \alpha_j)$, where $\alpha_1, \ldots, \alpha_n$ are the conjugates of an element generating L over K.
3. Find the discriminant of the ring of integers in $\mathbf{Q}(\sqrt[12]{1})$ (cf. Exercise 5 of section 2.3).
4. Show that the discriminant D of a finite extension of $\mathbf{Q}$ satisfies

$D \equiv 0,1 \pmod 4$. (*Hint*: $D = \det (u_i^{(j)})^2$, where $\{u_i\}$ is a basis and superscripts denote conjugates. Write $\det (u_i^{(j)}) = \alpha - \beta$, where α is the sum of the products corresponding to even permutations and β the rest. Observe that $\alpha + \beta$, $\alpha\beta$ are rational integers and $D = (\alpha - \beta)^2 = (\alpha + \beta)^2 - 4\alpha\beta$.)

5. Let L/K be a separable extension of degree n, $u_1, \ldots, u_n$ a basis and $P = (p_{ij})$, where $p_{ij} = u_i^{(j)}$ is the jth conjugate of u_i. If $A = (a_{ij})$ is the matrix of $a \in L$ in the regular representation: $u_i a = \Sigma a_{ij} u_j$, show that

$$P^{-1}AP = \operatorname{diag}(a^{(1)}, \ldots, a^{(n)}).$$

Deduce that $N(a)$ and $\mathrm{Tr}(a)$ are given by the determinant and trace of A, respectively.

6. Let $\mathfrak{o}$ be the ring of integers in an algebraic number field and let $\mathfrak{a}$ be an integral ideal in $\mathfrak{o}$. Given bases $u_1, \ldots, u_n$ of $\mathfrak{o}$ and $v_1, \ldots, v_n$ of $\mathfrak{a}$, if $v_i = \Sigma p_{ij} u_j$, show that $\det (p_{ij})$ is the number of elements in $\mathfrak{o}/\mathfrak{a}$ (and hence equal to $N(\mathfrak{a})$, cf. Exercise 3 of section 2.5). Deduce the following expression for the discriminant of $\mathfrak{a}$: $D(\mathfrak{a}) = N(\mathfrak{a})^2 D$, where D is the discriminant of $\mathfrak{o}$.

3

Global fields

An important property shared by algebraic number fields and algebraic function fields in one variable is the product formula and it suggests the following definition: a *global field* is a field with a family of absolute values satisfying the product formula. In section 3.2 we shall find that every global field is a finite algebraic extension of either **Q** or a rational function field $k(x)$, where k is a finite field. A small trick allows one to extend this result to the case of a perfect ground field k. This presentation follows Artin and Whaples [1945].

In any global field K we have a homomorphism θ from $K^{\times}$ to the divisor group and we go on to examine this map in greater detail in the number field case. We shall show in section 3.3 that its kernel is a finitely generated abelian group (Dirichlet unit theorem) and in section 3.4 we show that its cokernel, the divisor class group, is finite. The chapter begins with a look at quadratic extensions, to set the scene and illustrate the concepts introduced so far.

3.1 ALGEBRAIC NUMBER FIELDS

By an *algebraic number* one understands any solution of a polynomial equation with rational coefficients. Naturally the zero polynomial is excluded, so we can divide by the coefficient of the highest power, so as to obtain a monic equation

$$x^n + a_1 x^{n-1} + \ldots + a_n = 0, \quad \text{where } a_i \in \mathbf{Q}. \tag{1}$$

Any solution of (1) may be thought of as a complex number, but of course not every complex number is algebraic over **Q**. A complex number which is not algebraic is said to be *transcendental*. The existence of such 'transcendental' numbers is well known and it can be proved at three levels:

(i) It is easily checked that the set of all algebraic numbers is countable, whereas the set of all complex numbers is uncountable (this non-constructive proof goes back to Cantor).

(ii) Liouville has shown that loosely speaking, algebraic numbers cannot be approximated too closely by rational numbers. More precisely, for any algebraic number α satisfying an equation of degree $r > 1$ there exists a constant c such that $|\alpha - m/n| > cn^{-r}$ for all $m/n \in \mathbf{Q}$. This easily leads to transcendental numbers, e.g. $\Sigma 10^{-2^n} = 0.11010001\ldots$.

(iii) Specific real numbers such as e, π have been proved to be transcendental (by Hermite in 1873 and Lindemann in 1882 respectively).

A field consisting entirely of algebraic numbers is called an *algebraic number field*. We shall usually understand such a field to be finite-dimensional over $\mathbf{Q}$, since this is the case of interest to us here. Thus an algebraic number field will have a finite basis over $\mathbf{Q}$.

In order to study divisibility in an algebraic number field we have to define the notion of an algebraic integer. The natural way is to define an *algebraic integer* as a solution of an equation (1) when the a_i lie in $\mathbf{Z}$. In terms of the definition given in section 2.2, an algebraic integer is an element (of an extension field of $\mathbf{Q}$) which is integral over $\mathbf{Z}$. Since $\mathbf{Z}$ can be defined by a set S of prime divisors, the ring A of integers in a finite extension K of $\mathbf{Q}$ can be defined as the set S_K of all divisors of members of S. In particular, by Corollary 2.5.2, A is a Dedekind domain, though not necessarily principal.

Let us consider some examples. In $\mathbf{Q}(i)$, where i is a root of $x^2 + 1 = 0$, we have the integers $\alpha = a + bi$ $(a, b \in \mathbf{Z})$. They are called the *Gaussian integers*; $a + bi$ satisfies the equation

$$x^2 - 2ax + a^2 + b^2 = 0. \tag{2}$$

It is clear that every Gaussian integer is in fact integral over $\mathbf{Z}$. That conversely, every algebraic integer in $\mathbf{Q}(i)$ is of the form $a + bi$, where $a, b \in \mathbf{Z}$, requires a little proof, and is a good exercise at this point. For the answer see Theorem 1.3 below.

As a second example consider the equation $x^3 = 1$. This has the solution $\frac{1}{2}(-1 + \sqrt{-3})$, which is therefore also an algebraic integer, so the integers in $\mathbf{Q}(\sqrt{-3})$ include other numbers besides $\mathbf{Z}[\sqrt{-3}]$.

In practice the above definition leaves something to be desired: we merely know that α is an algebraic integer if it satisfies some monic equation with integer coefficients. But every algebraic number has a unique minimal equation which it satisfies, and it will be helpful to know that we can always use this minimal equation in testing for integrality:

Proposition 1.1

An algebraic number is an algebraic integer if and only if its minimal polynomial over $\mathbf{Q}$ has integral coefficients.

For the proof we need to recall a definition and a lemma. A polynomial over $\mathbf{Z}$ is called *primitive* if its coefficients have no common factor other than 1.

Lemma 1.2 (*Gauss's lemma*)
The product of primitive polynomials over **Z** *is primitive.*

Proof

Suppose that f, g are primitive, but that $h = fg$ is not; then all the coefficients of h are divisible by a prime p. Write $a \mapsto \bar{a}$ for the natural homomorphism from **Z** to $\mathbf{F}_p$, where $\mathbf{F}_p$ is the field of p elements. If we apply this homomorphism to f and g (which amounts to considering all coefficients mod p), we obtain $\bar{f}, \bar{g} \neq 0$, by primitivity, and since $\mathbf{F}_p[x]$ is an integral domain, $\bar{f}\bar{g} \neq 0$, but we also have $\bar{f}\bar{g} = \bar{h} = 0$, which is a contradiction.■

This proof uses the fact that **Z** is a unique factorization domain (UFD) and in fact the result holds for any UFD in place of **Z**.

We can now prove Proposition 1.1. Let α have the minimal polynomial g and suppose that α satisfies $f = 0$, where f is monic with integer coefficients; we have to show that g has integer coefficients. Since g is the minimal polynomial for α, we have $f = gh$ for some polynomial h (for if division by g left a non-zero remainder, this would give a polynomial of lower degree satisfied by α). Choose integers a, b such that ag, bh are primitive with integer coefficients. Then $abf = ag \,.\, bh$ is primitive, by Lemma 1.2, hence $ab = \pm 1$ and so $a = \pm 1$; therefore g has integer coefficients. Conversely, if g has integer coefficients, then α is integral, by definition.■

By the *degree* of an algebraic number one understands the degree of its minimal equation.

Given any algebraic number α, we can always find a positive integer m such that αm is an algebraic integer, as we saw in the proof of Proposition 2.2.4. Hence every algebraic number is of the form $\alpha = \beta m^{-1}$, where β is an algebraic integer and m a positive rational integer. It follows that for any algebraic number field we can take our basis (over **Q**) to consist of algebraic integers. Whether there is a **Z**-basis for the set of all algebraic integers is another question; fortunately this also has an affirmative answer, as we see by applying Theorem 2.6.2 and using the fact that **Z** is a principal ideal domain.

We now examine the simplest extensions, namely those of degree 2, called *quadratic extensions*, in more detail. Given a quadratic extension F of **Q**, F is two-dimensional and so is spanned by 1, α over **Q** for any $\alpha \in F \backslash \mathbf{Q}$. Here α satisfies a quadratic equation over $\mathbf{Q}: \alpha^2 + p\alpha + q = 0$. Replacing α by $\beta = \alpha + p/2$, we have $\beta^2 + q - (p/2)^2 = 0$, where $\beta^2 = b \in \mathbf{Q}$. By what was said earlier, we can take b to be an integer, which may be taken square free, for if $b = dr^2$, then we can replace β by β/r, where $(\beta/r)^2 = d$. This shows that every quadratic extension of **Q** is of the form

$$F = \mathbf{Q}(\sqrt{d}),$$

where d is a non-zero square free integer. Clearly we may also exclude the case $d = 1$, but it is possible for d to be negative.

For the special case of quadratic extensions it is easy to describe an explicit **Z**-basis:

Theorem 1.3
Any quadratic extension of $\mathbf{Q}$ is of the form $\mathbf{Q}(\sqrt{d})$, where d is a square free integer, not 0 or 1. The set of all algebraic integers in $\mathbf{Q}(\sqrt{d})$ forms a ring with the **Z**-*basis*

(i) $1, \sqrt{d}$ *if* $d \equiv 2$ or $3 \pmod 4$;

(ii) $1, (1+\sqrt{d})/2$ *if* $d \equiv 1 \pmod 4$.

Proof
We have just seen that the first sentence holds. That the integers in $\mathbf{Q}(\sqrt{d})$ form a ring follows from Corollary 2.5.2, but it can also be verified directly, using the explicit basis to be found.

Let us now examine when $\alpha = a + b\sqrt{d}$ $(a, b \in \mathbf{Q})$ is an integer. If $b = 0$, a is an algebraic integer if and only if $x - a$ has integer coefficients. Thus a rational number is an algebraic integer if and only if it is a rational integer (as is to be expected from section 2.5). Suppose now that $b \neq 0$; then the minimal equation for α is

$$x^2 - 2ax + a^2 - b^2d = 0.$$

Hence α is an algebraic integer if and only if

$$2a \in \mathbf{Z}, \qquad a^2 - b^2d \in \mathbf{Z}. \tag{3}$$

This condition certainly holds when $a, b \in \mathbf{Z}$; we have to find when there are other cases. Write $2a = a'$, $2b = b'$; then

$$4(a^2 - b^2d) = a'^2 - b'^2d,$$

hence (3) holds precisely when

$$a' \in \mathbf{Z}, \qquad a'^2 - b'^2d \in 4\mathbf{Z}. \tag{4}$$

It follows that $a', b'^2d \in \mathbf{Z}$. We can write $b' = m/n$, where m, n are coprime integers and $n > 0$. Since b'^2d is integral, we have $n^2 \mid m^2d$, but d is square free, therefore $n|m$ and so $b' \in \mathbf{Z}$. Thus a', b' are integers and by (4),

$$a'^2 \equiv b'^2d \pmod 4. \tag{5}$$

We now distinguish several cases:

(i) $d \equiv 2$ or $3 \pmod 4$. Suppose that b' is odd; then $b'^2 \equiv 1 \pmod 4$, hence $a'^2 \equiv d \pmod 4$, but this is impossible because the square a'^2 is $\equiv 0$ or $1 \pmod 4$. Hence b' is even and by (5), so is a'. This shows the integers in $\mathbf{Q}(\sqrt{d})$ to be of the form $a + b\sqrt{d}$, where $a, b \in \mathbf{Z}$.

(ii) $d \equiv 1 \pmod 4$. Then $a'^2 \equiv b'^2 \pmod 4$ by (5), so a', b' are either both even or both odd, and in each case (4) is satisfied. So the integers in $\mathbf{Q}(\sqrt{d})$ have the form

$$r + s\sqrt{d} \quad \text{or} \quad r + \frac{1}{2} + \left(s + \frac{1}{2}\right)\sqrt{d} \quad (r, s \in \mathbf{Z}),$$

i.e. $a + b(1 + \sqrt{d})/2$, where $a, b \in \mathbf{Z}$.

(iii) $d \equiv 0 \pmod 4$. This means that d is divisible by $4 = 2^2$, a case that was excluded.

Now it is easily verified that in each case (i), (ii) the sum and product of two numbers are again of the same form; this shows again that the set of all algebraic integers in $\mathbf{Q}(\sqrt{d})$ is a ring.■

We note that when α is of degree 2 over $\mathbf{Q}$, the norm $N(\alpha)$ is the constant term of the minimal equation for α and so α is a unit if and only if $N(\alpha) = \pm 1$. If $\alpha = a + b\sqrt{d}$, then $N(\alpha) = a^2 - b^2 d$, while for $\alpha = a + b(1 + \sqrt{d})/2$,

$$N(\alpha) = (a + b/2)^2 - (b/2)^2 d = a^2 + ab - b^2(d - 1)/4.$$

From section 2.5 we know that the algebraic integers in $\mathbf{Q}(\sqrt{d})$ form a Dedekind domain and the question arises under what circumstances this ring is a principal ideal domain (PID), or equivalently (by Proposition 2.5.4) a unique factorization domain (UFD). For the moment we shall confine ourselves to giving examples of non-principal domains. Let us take the ring A of integers in $\mathbf{Q}(\sqrt{-5})$; since $-5 \equiv 3 \pmod 4$, the integers are all of the form $a + b\sqrt{-5}$ $(a, b \in \mathbf{Z})$. Consider the equation

$$6 = 2.3 = (1 + \sqrt{-5})(1 - \sqrt{-5}). \tag{6}$$

We claim that 2, 3, $1 \pm \sqrt{-5}$ are all atoms, so that (6) gives two essentially different factorizations of 6. Consider $1 + \sqrt{-5}$; we have $N(1 + \sqrt{-5}) = 6$. Now if $1 + \sqrt{-5} = \alpha\beta$, then

$$6 = N(\alpha\beta) = N(\alpha)N(\beta),$$

so if α, β are non-units, then $N(\alpha)$, $N(\beta)$ must be 2 and 3, say $N(\alpha) = 2$, $N(\beta) = 3$. But the equation $x^2 + 5y^2 = 2$ or 3 has no integer solutions. This shows $1 + \sqrt{-5}$ to be an atom; similarly for $1 - \sqrt{-5}$ and 2, 3. Moreover, the only units in A are ± 1, for the equation $x^2 + 5y^2 = 1$ has only the solutions $(x, y) = (\pm 1, 0)$; this makes it clear that the atoms in the factorization (6) are not associated.

If we now consider ideals in A, we have

$$(6) = (2)(3) = (1 + \sqrt{-5})(1 - \sqrt{-5}),$$

and this can be further decomposed: we have $(6) = \mathfrak{p}_1\mathfrak{p}_2\mathfrak{p}_3\mathfrak{p}_4$, where $\mathfrak{p}_1 = (2, 1 + \sqrt{-5})$, $\mathfrak{p}_2 = (2, 1 - \sqrt{-5})$, $\mathfrak{p}_3 = (3, 1 + \sqrt{-5})$, $\mathfrak{p}_4 = (3, 1 - \sqrt{-5})$. Now $2 = \mathfrak{p}_1\mathfrak{p}_2$ is no longer a prime ideal. That the $\mathfrak{p}_i$ are prime ideals can be seen by taking norms, e.g. $N(\mathfrak{p}_1) = (4, 6) = (2)$, $N(\mathfrak{p}_3) = (9, 6) = (3)$, etc.

As another example we take $d = -3$. It can be shown that the ring of all algebraic integers in $\mathbf{Q}(\sqrt{-3})$ is a UFD (using the Euclidean algorithm), but the ring

$$\mathbf{Z}[\sqrt{-3}] = \{a + b\sqrt{(-3)} \mid a, b \in \mathbf{Z}\}$$

is not a UFD, as the equation

$$4 = 2.2 = (1 + \sqrt{-3})(1 - \sqrt{-3})$$

shows, by a similar method to that used on (6). In any case this ring cannot be a UFD since it is not integrally closed in $\mathbf{Q}(\sqrt{-3})$.

As a third example consider the ring $A = \mathbf{C}[x, \sqrt{(1 - x^2)}]$, where x is an indeterminate. Writing $y = \sqrt{(1 - x^2)}$, we have

$$y^2 = (1 + x)(1 - x)$$

and it is again easily checked that $1 \pm x$, y are atoms. This shows that A is not a UFD, although it is a Dedekind domain, as the integral closure of $\mathbf{C}[x]$ in $\mathbf{C}(x, y)$.

To examine factorizations in quadratic extensions, at least in the principal case, we shall need some information on primes:

Theorem 1.4

Let A be the ring of algebraic integers in a quadratic extension of $\mathbf{Q}$ and suppose that A is a unique factorization domain. Then each prime of A divides just one rational prime, and if p is the rational prime divisible by the prime π, then either π is associated to p, so that p is prime in A, or $p = |N(\pi)|$.

Proof

Let π be a prime in A and π^* its conjugate, so that $\pi\pi^* = N(\pi) \in \mathbf{Z}$. If the least positive rational multiple of π is m, and $m = m'm''$, then $\pi \mid m'm''$, hence $\pi \mid m'$ or $\pi \mid m''$, because π is prime. This shows m to be prime. If π divides two non-associated primes, say p', p'', then π divides $up' + vp''$ for $u, v \in \mathbf{Z}$, and for suitable u, v this is 1, so $\pi \mid 1$, a contradiction. Hence π divides exactly one rational prime, p say.

We now have $\pi \mid p$, say $p = \pi\mu$; hence $N(\pi)N(\mu) = p^2$. Either $N(\mu) = \pm 1$, so μ is a unit and π is associated to p, or $|N(\mu)| = p$. Since $|N(\pi)| > 1$, there are no other possibilities.■

This result can be used to obtain factorizations in rings of algebraic integers which are UFDs. We shall illustrate this by the simplest case, the Gaussian integers. For any ring A of algebraic integers which is a UFD we define a rational prime p to be *inert* if p stays prime in A, *decomposed* if it splits into distinct factors and *ramified* if there are repeated factors in the complete factorization of p in A.

In working out our example we shall need the fact that the multiplicative group of a finite field is cyclic; we briefly recall the proof. Let F be a field of $r+1$ elements; then $F^\times$ is a group of order r. By the basis theorem for abelian groups $F^\times$ is a direct product of cyclic prime power groups. If all the primes occurring are distinct, then this direct product is itself a cyclic group, so if $F^\times$ is not cyclic, it contains, for some prime p, a non-cyclic subgroup of order p^2. This means that the equation $x^p = 1$ has at least p^2 solutions, which is impossible over a field.

The ring of Gaussian integers is $A = \mathbf{Z}[i]$, where $i^2 = -1$. It is seen to be a UFD by the Euclidean algorithm, which holds practically unchanged. Thus, given $\alpha, \beta \in A, \beta \neq 0$, we need to find γ, δ such that $\alpha = \beta\gamma + \delta$, $N(\delta) < N(\beta)$. We can restate this for $\xi = \alpha\beta^{-1}$ as follows:

Given $\xi \in \mathbf{Q}(i)$, there exists an integer γ in $\mathbf{Q}(i)$ such that

$$N(\xi - \gamma) < 1. \tag{7}$$

If $\xi = x + yi$ $(x, y \in \mathbf{Q})$, we can find rational integers a, b at most $\frac{1}{2}$ distant from x, y respectively. Thus $|x - a| \leq \frac{1}{2}$, $|y - b| \leq \frac{1}{2}$, and so, writing $\gamma = a + bi$, we have

$$N(\xi - \gamma) = (x-a)^2 + (y-b)^2 \leq \frac{1}{4} + \frac{1}{4} < 1.$$

This establishes (7) and it shows that $A = \mathbf{Z}[i]$ satisfies the Euclidean algorithm relative to the norm function. Now a familiar argument (cf. A. 1, 10.5) shows that A is a PID and a UFD.

Let us take a prime π in A and write $\pi = a + bi$. Then $N(\pi) = a^2 + b^2$, so $p = N(\pi)$ means that $p = a^2 + b^2$. Therefore by Theorem 1.4, p is inert if and only if the equation $p = x^2 + y^2$ has no solution in integers x, y. We distinguish three cases:

(i) $p \equiv 3 \pmod 4$. Then $p \neq a^2 + b^2$, since a sum of two squares is $\equiv 0$, 1 or 2 (mod 4). Hence p is inert in this case.
(ii) $p \equiv 2 \pmod 4$. This means that $p = 2 = (1+i)^2(-i)$, so 2 is ramified.
(iii) $p \equiv 1 \pmod 4$. We claim that p decomposes. In the first place the congruence

$$x^2 + 1 \equiv 0 \pmod p \tag{8}$$

has a solution. For if $p = 4n + 1$ say, then $F_p^\times$ is cyclic of order $4n$; if u is a generator, then $v = u^{2n}$ satisfies $v^2 \equiv 1$, $v \not\equiv 1 \pmod p$, hence $v \equiv -1 \pmod p$, and so $x = u^n$ is a solution of (8).

Writing $c = u^n$, we have $p \mid c^2 + 1 = (c+i)(c-i)$. If p were prime in A, it would divide $c + i$ or $c - i$, but p is odd, so $c/p \pm i/p$ is not integral and hence this cannot happen. Thus p is not prime, say $p = \pi\bar{\pi}$, where $\pi = a + bi$, $\bar{\pi} = a - bi$, $p = a^2 + b^2$ and the factors are not associated. For the units in A are

$\pm 1, \pm i$ and clearly $a - bi \neq \pm (a + bi)$; if $a - bi = -i(a + bi)$, say, then $a = b$, so $p = 2a^2 \equiv 0 \pmod 2$, which contradicts the fact that p is odd.

We can summarize the result as follows.

Theorem 1.5

The ring of Gaussian integers $\mathbf{Z}[i]$ *is a unique factorization domain, whose primes are as follows:*

(i) *Any rational prime* $p \equiv 3 \pmod 4$ *is inert, i.e. remains prime in* $\mathbf{Z}[i]$;

(ii) *any rational prime* $p \equiv 1 \pmod 4$ *decomposes*:

$$p = a^2 + b^2 = (a + bi)(a - bi);$$

(iii) 2 *is ramified*: $2 = (1 + i)^2(-i)$.■

As a consequence we have a result going back to Fermat, fully proved by Euler:

Corollary 1.6

A rational prime p *can be written as a sum of two squares if and only if* $p \equiv 1$ *or* $2 \pmod 4$.■

In connexion with Theorem 1.5 it is instructive to take the p-adic valuation on $\mathbf{Q}$ and look for its extensions to $\mathbf{Q}(i)$. Since this is a Galois extension, we have $efr = 2$, by Theorem 2.3.4.

If $p \equiv 3 \pmod 4$, then $x^2 + 1 = 0$ remains irreducible over the residue class field $\mathbf{F}_p$, so the residue class field of $\mathbf{Q}(i)$ is a quadratic extension of $\mathbf{F}_p$; here $f = 2$, $e = r = 1$.

If $p \equiv 1 \pmod 4$, then $x^2 + 1 = 0$ becomes reducible over $\mathbf{F}_p$, so the residue class field remains unchanged, as is the value group. Hence $e = f = 1$ and so $r = 2$.

If $p = 2$, the value group is extended, because 2 becomes a square. Thus $e = 2, f = r = 1$.

Exercises

1. Verify that an algebraic number is a unit if and only if the coefficients of its minimal polynomial are integers and the constant term is ± 1.
2. Show that the number of algebraic integers of degree n whose conjugates are all of absolute value 1 is finite (for a given n), and find a bound in terms of n. If this holds for a number α and all its powers, show that α is a root of 1.
3. Show that an algebraic integer whose conjugates are all of absolute value 1 is a root of 1. Does the conclusion hold for any algebraic integer of absolute value 1? (*Hint:* Write down an equation for a unit with a coefficient which is an algebraic integer with insoluble Galois group.)
4. Show that any root of a monic equation with algebraic integer coefficients

is an algebraic integer.

5. Verify that the ring of integers in $\mathbf{Q}(\sqrt{d})$ is Euclidean for $d = 2, 3, 5$ and for $d = -1, -2, -3, -5, -7, -11$.
6. Show that the units in $\mathbf{Q}(\rho)$, where ρ has the minimal equation $x^2 + x + 1 = 0$, form a cyclic group of order 6. Show further that any prime congruent to 2 (mod 3) is inert, any prime congruent to 1 (mod 3) decomposes and 3 is ramified.
7. Let K be an extension of degree n of $\mathbf{Q}$ and let v be a valuation on $\mathbf{Q}$ with a residue class field of at least n elements. Show that the ring of integers in K satisfies the Euclidean algorithm with respect to the function $\phi(\alpha) = 2^{v(N(\alpha))}$ (Borevich–Shafarevich).
8. For any ideals $\mathfrak{a}$, $\mathfrak{b}$, $\mathfrak{c}$ in a ring verify the identity:

$$(\mathfrak{a} + \mathfrak{b} + \mathfrak{c})(\mathfrak{b}\mathfrak{c} + \mathfrak{c}\mathfrak{a} + \mathfrak{a}\mathfrak{b}) = (\mathfrak{b} + \mathfrak{c})(\mathfrak{c} + \mathfrak{a})(\mathfrak{a} + \mathfrak{b}).$$

   Deduce that an integral domain is a Dedekind domain whenever it is Noetherian and every two-generator ideal is invertible (Dedekind).
9. Let K be a quadratic extension of $\mathbf{Q}$, A its ring of integers and write $\alpha \mapsto \alpha'$ for the automorphism of $K/\mathbf{Q}$. Given $\alpha, \beta \in A$ and $n \in \mathbf{Z}$, show that if n divides $\alpha\alpha'$, $\beta\beta'$, $\alpha\beta' + \alpha'\beta$, then $n \mid \alpha'\beta$. Deduce that for any ideal $\mathfrak{a}$ in A, $\mathfrak{a}\mathfrak{a}' = nA$ for some $n \in \mathbf{Z}$, hence show that A is Dedekind. (*Hint:* Write $\gamma = \alpha'\beta/n$ and verify that γ is an algebraic integer. If $\mathfrak{a} = (\alpha, \beta)$, show that

$$\mathfrak{a}\mathfrak{a}' \supseteq (\alpha\alpha', \beta\beta', \alpha\beta' + \alpha'\beta) \supseteq (n)$$

   and use the first part. For the final part use Exercise 8.)
10. (Dedekind) Let K be an algebraic number field with ring of integers generated by a single algebraic integer α with minimal polynomial f. Show that any rational prime p factorizes in K as

$$p = \mathfrak{p}_1^{e_1} \ldots \mathfrak{p}_r^{e_r}, \quad \text{where} \quad \mathfrak{p}_i = (f_i(\alpha), p)$$

   and $f \equiv f_1^{e_1} \ldots f_r^{e_r}$ (mod p) is a complete factorization of f (mod p). Use this result to rederive the decomposition of primes in the ring of Gaussian integers (Theorem 1.5).

3.2 THE PRODUCT FORMULA

The fields $\mathbf{Q}$, $k(x)$ and their rings of integers have a further remarkable property which is actually enough to characterize them and their finite algebraic extensions.

Consider first $\mathbf{Q}$ and its subring $\mathbf{Z}$. Let v_p be the normalized valuation for the prime p, and define the p-adic absolute value by the equation

$$|a|_p = p^{-v_p(a)},$$

while the usual absolute value on $\mathbf{Q}$ is written $|a|_\infty$. Then, since

$a = \pm \Pi p^{v_p(a)}$ we have the *product formula*:

$$\prod_{p,\infty} |a|_p = 1 \quad \text{for all } a \in \mathbf{Q}^{\times}. \tag{1}$$

This holds for $a \in \mathbf{Z}^{\times}$ in the first place and hence for all $a \in \mathbf{Q}^{\times}$. It is the only relation between all the absolute values on $\mathbf{Q}$, for if we had another relation $\Pi |a|_p^{\alpha_p} = 1$, independent of (1), then on dividing this by the relation (1) raised to the power α_∞, we would find

$$\Pi |a|_p^{\beta_p} = 1, \quad \text{where } \beta_\infty = 0. \tag{2}$$

If $\beta_p \neq 0$ for some p, say $\beta_p > 0$, and we take $a = p^N$, then for large N, $|a|_p^{\beta_p} < 1$ while $|a|_q = 1$ for all $q \neq p$, and this contradicts (2). Thus (1) is the only relation between the absolute values on $\mathbf{Q}$.

A similar product formula exists for $k(x)$. For each irreducible polynomial p of degree δ_p let us define the p-adic absolute value on $k(x)$ as

$$|a|_p = c^{-\delta_p v_p(a)},$$

where c is fixed, $c > 1$. Corresponding to x^{-1} we have the further place at ∞, given by $-\deg a$, as we saw in the proof of Proposition 1.4.4. If $a = \Pi p^{v_p}$, then $\deg a = \Sigma\, \delta_p v_p$, hence on defining $|a|_\infty$ as $c^{\deg a}$, we have

$$\Pi |a|_p = c^{-\Sigma \delta_p v_p + \deg a} = 1. \tag{3}$$

So we again have a product formula and as before we can verify that it is the only relation of its kind for $k(x)$.

We remark that in each case we have a family S of absolute values $| \; |_p$, archimedean or non-archimedean, such that the product formula holds for all $p \in S$, but if we omit a single one (the archimedean absolute value in the case of $\mathbf{Q}$), the remaining set has the strong approximation property and so defines a Dedekind domain. The members of S will again be called *places*.

Now let K be any field and let S be a set of places on K for which a product formula holds. This means in detail that for any $a \in K^{\times}$ we have $|a|_{\mathfrak{p}} = 1$ for almost all $\mathfrak{p} \in S$ and

$$\prod_{\mathfrak{p} \in S} |a|_{\mathfrak{p}} = 1. \tag{4}$$

Here we have chosen the normalization for each absolute value so that its exponent in (4) is 1; clearly this represents no loss of generality. However, the strong approximation property of section 2.4 is not assumed. In fact, it will not usually hold because the set

$$A = \{a \in K \mid |a|_{\mathfrak{p}} \leqslant 1 \text{ for all } \mathfrak{p} \in S\}$$

will be a subring only if all the absolute values are non-archimedean, and even then it may not have K as its field of fractions (in that case A itself will be a field, cf. the proof of Proposition 2.3 below).

We remark that although S may include archimedean absolute values, their number must be finite, because such an absolute value is characterized by $|2|>1$, by Proposition 1.1.2.

As a first result we note that the product formula goes over to finite algebraic extensions.

Proposition 2.1
Let K be a field with a set S of places satisfying a product formula, and let L be a finite separable algebraic extension of K. If S_L is the set of all places on L extending the places in S, then the members of S_L again satisfy a product formula.

Proof
It will be convenient to take the product formula in K in additive notation as

$$\sum v_{\mathfrak{p}}(a) = 0 \quad \text{for } a \in K^{\times},$$

where $v_{\mathfrak{p}}$ is a rank 1 valuation or $-\log$ of an absolute value.

Given $L \supseteq K$ as stated and $\mathfrak{p} \in S$, let $\mathfrak{P}_1, \ldots, \mathfrak{P}_r$ be the extensions of $\mathfrak{p}$ to L. We have seen that they are given by decomposing the tensor product of L with the completion $\hat{K}$, as in Theorem 2.3.3:

$$L \otimes \hat{K} \cong L_1 \times \ldots \times L_r.$$

By Corollary 2.3.2 we have

$$N_{L/K}(\alpha) = \Pi\, N_{L_i/\hat{K}}(\alpha),.$$

and L_i is complete under $\mathfrak{P}_i$, so the corresponding extension $w_{\mathfrak{P}_i}$ of $v_{\mathfrak{p}}$ is given by

$$w_{\mathfrak{P}_i}(\alpha) = [L_i : \hat{K}]^{-1} v_{\mathfrak{p}}(N_{L_i/\hat{K}}(\alpha)).$$

Hence

$$\begin{aligned}\sum [L_i : \hat{K}]\, w_{\mathfrak{P}_i}(\alpha) &= \sum v_{\mathfrak{p}}(N_{L_i/\hat{K}}(\alpha)) \\ &= \sum v_{\mathfrak{p}}(N_{L/K}(\alpha)) = 0. \blacksquare\end{aligned}$$

For our next result we will have to count (or at least estimate) the number of integral elements in a certain region. The classical theorem of Minkowski does this for symmetric convex regions in algebraic number fields. We shall only need it for parallelotopes, but we want to cover the algebraic function case as well. We thus need two concepts, *volume* and *order* (i.e. number of elements) of a parallelotope.

Let K be a field with a family S of absolute values satisfying a product formula. By a *parallelotope* $\mathfrak{A}$ in K we understand a function which associates with each $\mathfrak{p} \in S$ an element $|\mathfrak{A}|_{\mathfrak{p}}$ in the value group of $\mathfrak{p}$ (in multiplicative notation), such that $|\mathfrak{A}|_\mathfrak{p} = 1$ for almost all $\mathfrak{p} \in S$. The *volume* of $\mathfrak{A}$ is defined as

$$\mu(\mathfrak{A}) = \prod |\mathfrak{A}|_\mathfrak{p}.$$

Parallelotopes can be multiplied by the rule

$$|\mathfrak{A}\mathfrak{B}|_\mathfrak{p} = |\mathfrak{A}|_\mathfrak{p} \,.\, |\mathfrak{B}|_\mathfrak{p}.$$

It follows immediately that

$$\mu(\mathfrak{A}\mathfrak{B}) = \mu(\mathfrak{A}) \,.\, \mu(\mathfrak{B}). \tag{5}$$

We see that a parallelotope is very much like a divisor (and hence a fractional ideal), except that the family S here is bigger and may include archimedean values. For example, on $\mathbf{Q}$ a parallelotope is a divisor together with a real positive number $|\mathfrak{A}|_\infty$.

Every element $c \in K^\times$ defines a parallelotope by the rule $c \mapsto |c|_p$ and the product formula just states that $\mu(c) = 1$. Hence, bearing in mind (5), we find that

$$\mu(c\mathfrak{A}) = \mu(\mathfrak{A}) \quad \text{for } c \in K^\times. \tag{6}$$

With each parallelotope $\mathfrak{A}$ we associate a subset $L(\mathfrak{A})$ of K defined by

$$L(\mathfrak{A}) = \{x \in K \,|\, |x|_\mathfrak{p} \leqslant |\mathfrak{A}|_\mathfrak{p} \text{ for all } \mathfrak{p} \in S\}. \tag{7}$$

To illustrate this definition we look at the cases $K = \mathbf{Q}, k(x)$.

(i) $K = \mathbf{Q}$. A parallelotope $\mathfrak{A}$ is specified by

$$|\mathfrak{A}|_p = p^{-\alpha_p} \quad (\alpha_p \in \mathbf{Z}), \qquad |\mathfrak{A}|_\infty = \gamma_0, \text{ where } \gamma_0 \in \mathbf{R}.$$

Here $\mu(\mathfrak{A}) = \gamma_0/c$, where $c = \Pi p^{\alpha_p}$. Consider $L(\mathfrak{A}) : x \in L(\mathfrak{A})$ means that $|x|_p \leqslant p^{-\alpha_p}$, i.e. $c \,|\, x$, in other words, $xc^{-1} \in \mathbf{Z}$, and $|x|_\infty \leqslant \gamma_0$. So $L(\mathfrak{A})$ consists of all the multiples of c in the region $|x|_\infty \leqslant \gamma_0$, i.e. all nc, where $n \in \mathbf{Z}, |n| \leqslant \gamma_0/c$. The number of such elements is $2[\gamma_0/c] + 1$, and if we denote this number by $\lambda(\mathfrak{A})$, then we have

$$2\mu(\mathfrak{A}) - 1 \leqslant \lambda(\mathfrak{A}) \leqslant 2\mu(\mathfrak{A}) + 1.$$

This means that $\lambda(\mathfrak{A})$, the number of points in $L(\mathfrak{A})$, is of the same order of magnitude as $\mu(\mathfrak{A})$.

(ii) $K = k(x)$. Here we have only non-archimedean values. A parallelotope $\mathfrak{A}$ is specified by

$$|\mathfrak{A}|_\mathfrak{p} = c^{-\delta_\mathfrak{p}\alpha_\mathfrak{p}}, \qquad |\mathfrak{A}|_\infty = c^{\alpha_\infty} \quad \text{where } \alpha_\mathfrak{p}, \alpha_\infty \in \mathbf{Z}.$$

Here $\mathfrak{p}$ ranges over the irreducible polynomials p and $\delta_p = \deg p$. Given $u \in k(x)$, if u lies in $L(\mathfrak{A})$, we have $|u|_\mathfrak{p} \leqslant |\mathfrak{A}|_\mathfrak{p}$ for all $\mathfrak{p}$, i.e. u is divisible by $\phi = \Pi p^{\alpha_\mathfrak{p}}$, say $u = \phi f$, where f is a polynomial. Further,

$|u|_\infty \leqslant |\mathfrak{A}|_\infty$ means that $\deg u \leqslant \alpha_\infty$; thus

$$\deg f \leqslant \alpha_\infty - \deg \phi = \alpha_\infty - \sum \delta_\mathfrak{p}\alpha_\mathfrak{p}.$$

This shows that $u \in L(\mathfrak{A})$ if and only if $u = \phi f$, where f is a polynomial of degree at most $\log_c \mu(\mathfrak{A})$.

If we assume k to be finite with q elements, then the number of polynomials of degree at most r is q^{r+1}, so in this case the number of elements in $L(\mathfrak{A})$ satisfies

$$\lambda(\mathfrak{A}) = q^{r+1}, \quad \text{where } \mu(\mathfrak{A}) = c^r.$$

Here the constant c is at our disposal; let us put $c = q$. Then

$$\lambda(\mathfrak{A}) = q \,.\, \mu(\mathfrak{A}). \tag{8}$$

Thus $\lambda(\mathfrak{A})$ and $\mu(\mathfrak{A})$ are again of the same order of magnitude.

We now return to the case of an arbitrary field K with a set S of places satisfying the product formula. We shall take the values to be normalized as follows:

1. $\mathfrak{p}$ archimedean, with completion $\mathbf{R}$: $|x|_\mathfrak{p} = |x|$, the usual absolute value;
2. $\mathfrak{p}$ archimedean, with completion $\mathbf{C}$: $|x|_\mathfrak{p} = |x|^2$;
3. $\mathfrak{p}$ non-archimedean principal, with normalized valuation $v_\mathfrak{p}$; assuming the residue class field to be finite, with q elements, we write $|x|_\mathfrak{p} = q^{-v_\mathfrak{p}(x)}$.

As we shall see, the number of elements in $L(\mathfrak{A})$, for a parallelotope $\mathfrak{A}$, is again finite; it will be called the *order* of $\mathfrak{A}$ and denoted by $\lambda(\mathfrak{A})$. The scope of this result can still be extended to cover the case of infinite residue class fields. Suppose that the residue class fields are finite extensions of a fixed field k. We shall assign a number $q > 1$ as *order* to k and for a place $\mathfrak{p}$ with residue class field E, of degree r over k, define its order as

$$N\mathfrak{p} = q^r.$$

Further, we put $|\pi|_\mathfrak{p} = N\mathfrak{p}^{-1}$ for a uniformizer π; using this notion of order of a residue class field we can define the *order* $\lambda(\mathfrak{A})$ of a parallelotope $\mathfrak{A}$ as before. We remark that for any $c \in K^\times$ the map $\mathfrak{A} \mapsto c\mathfrak{A}$ induces a bijection $L(\mathfrak{A}) \to L(c\mathfrak{A})$; it follows that

$$\lambda(c\mathfrak{A}) = \lambda(\mathfrak{A}) \quad \text{for } c \in K^\times. \tag{9}$$

We now define a *global field* to be a field with a set of places satisfying a product formula with the above normalization. The preceding examples show $\mathbf{Q}$ and $k(x)$ to be global fields, also called *rational* global fields; with the above definition of order of k, this holds even for an infinite ground field k.

By Proposition 2.1, any separable extension of a global field is again a global field; to be precise, we have to make sure that the product formula holds for the extension with this normalization, but this is easily checked. We have

$$w_i(\alpha) = (1/n_i)v_\mathfrak{p}(N_{L_i/\hat{K}}(\alpha)),$$

where w_i is an extension of $v_\mathfrak{p}$ to L and $n_i = [L_i : \hat{K}]$. If the corresponding normalized valuation is v_i and $n_i = e_i f_i$ as before, then

$$v_i(\alpha) = e_i w_i(\alpha) = (1/f_i)v_\mathfrak{p}(N_{L_i/\hat{K}}(\alpha)),$$

and by (8) of section 2.5, $N_{L_i/\hat{K}}(\mathfrak{P}_i) = \mathfrak{p}^{f_i}$, hence

$$N(\mathfrak{P}_i) = (N\mathfrak{p})^{f_i}.$$

So for $\alpha \in L$ we have $\Sigma f_i v_i(\alpha) = v_\mathfrak{p}(N((\alpha))$, and hence

$$\prod |\alpha|_{\mathfrak{P}_i} = \prod (N\mathfrak{P}_i)^{-v_i(\alpha)} = (N\mathfrak{p})^{-\Sigma f_i v_i(\alpha)} = (N\mathfrak{p})^{-v_\mathfrak{p}(N(\alpha))}.$$

A corresponding calculation shows that the norm at each archimedean place is preserved and so

$$\prod |\alpha|_\mathfrak{P} = \prod |N(\alpha)|_\mathfrak{p} = 1. \blacksquare$$

The above examples showed $\lambda(\mathfrak{A})$ to be of the same order of magnitude as $\mu(\mathfrak{A})$, and we shall find that this holds generally. It is fairly straightforward to prove that the order is bounded by $\mu(\mathfrak{A})$ and we shall do so now, leaving the reverse inequality until section 3.3.

Proposition 2.2
Let K be a global field. Then there exists a constant M such that

$$\lambda(\mathfrak{A}) \leqslant \max\{M \,.\, \mu(\mathfrak{A}), 1\} \quad \textit{for all parallelotopes } \mathfrak{A}. \tag{10}$$

Here the 1 on the right of (10) is needed because $\lambda(\mathfrak{A})$ assumes discrete values with least value 1.

Proof
We have seen that there are only finitely many archimedean absolute values. Let r be their number, fix a non-archimedean place $\mathfrak{q}$ and put

$$M = 4^r \,.\, N\mathfrak{q}. \tag{11}$$

Given any parallelotope $\mathfrak{A}$, we know that $\mathfrak{A}$ and $c\mathfrak{A}$, for $c \in K^\times$, have the same order and the same volume, so we may in proving (10) replace $\mathfrak{A}$ by $c\mathfrak{A}$ for any $c \in K^\times$. We choose c so that

$$|\mathfrak{A}|_\mathfrak{q} = 1, \tag{12}$$

bearing in mind that $\mathfrak{q}$ is a fixed place. If $L(\mathfrak{A}) = 0$, (10) is clear, so we can take a subset X of $L(\mathfrak{A})$ of order $N_0 > 1$. Writing $\mathfrak{o}$, $\mathfrak{q}$ for the valuation ring $\mathfrak{o}_\mathfrak{q}$ and its maximal ideal, we have

$$X \subseteq L(\mathfrak{A}) \subseteq \mathfrak{o},$$

and

$$\mathfrak{o} \supseteq \mathfrak{q} \supseteq \mathfrak{q}^2 \supseteq \ldots,$$

and we can choose an integer n such that

$$(N\mathfrak{q})^n < N_0 \leqslant (N\mathfrak{q})^{n+1}. \tag{13}$$

By hypothesis the residue class field $\mathfrak{o}/\mathfrak{q}$ has finite order $N\mathfrak{q}$. Now $\mathfrak{q}$ is a principal ideal in $\mathfrak{o}$, therefore

$$\mathfrak{o}/\mathfrak{q}^{n-1} \cong \mathfrak{o}/\mathfrak{q}^n/\mathfrak{q}^{n-1}/\mathfrak{q}^n.$$

Denoting the order of a residue class ring again by λ, we have

$$\lambda(\mathfrak{o}/\mathfrak{q}^{n-1}) = \lambda(\mathfrak{o}/\mathfrak{q}^n)/\lambda(\mathfrak{o}/\mathfrak{q}),$$

and by an easy induction we find

$$\lambda(\mathfrak{o}/\mathfrak{q}^n) = (N\mathfrak{q})^n.$$

By (13), $\lambda(X) > (N\mathfrak{q})^n$, so there exist $\alpha, \beta \in X$ such that α, β are distinct but congruent mod $\mathfrak{q}^n$:

$$|\alpha - \beta|_{\mathfrak{q}} \leqslant |\mathfrak{q}^n|_{\mathfrak{q}} = (N\mathfrak{q})^{-n} \leqslant N\mathfrak{q}/N_0. \tag{14}$$

Since $\alpha, \beta \in L(\mathfrak{A})$, we have

$$|\alpha - \beta|_{\mathfrak{p}} \leqslant \begin{cases} |\mathfrak{A}|_{\mathfrak{p}} & \text{if } \mathfrak{p} \text{ is non–archimedean and } \mathfrak{p} \neq \mathfrak{q}, \\ 4|\mathfrak{A}|_{\mathfrak{p}} & \text{if } \mathfrak{p} \text{ is archimedean.} \end{cases}$$

Using the product formula and (14), we find

$$1 = \prod |\alpha - \beta|_{\mathfrak{p}} \leqslant 4^r \Big(\prod_{\mathfrak{p} \neq \mathfrak{q}} |\mathfrak{A}|_{\mathfrak{p}} \Big) \frac{(N\mathfrak{q})}{N_0},$$

but $|\mathfrak{A}|_{\mathfrak{q}} = 1$; hence

$$1 \leqslant \frac{4^r \mu(\mathfrak{A}) \,.\, N\mathfrak{q}}{N_0},$$

and so $N_0 \leqslant 4^r \mu(\mathfrak{A}) \,.\, N\mathfrak{q}$. Here the right-hand side is independent of X, so $\lambda(\mathfrak{A})$ is finite, and taking $X = L(\mathfrak{A})$, we obtain the bound $M \,.\, \mu(\mathfrak{A})$ for $\lambda(\mathfrak{A})$, as claimed.■

We shall use this estimate for $\lambda(\mathfrak{A})$ to identify global fields more precisely:

Proposition 2.3
Every global field is a finite algebraic extension either of $\boldsymbol{Q}$ *or of the rational function field* $k(x)$.

Proof

Let K be a global field. By hypothesis we have a product formula on K, with the normalization as described on p.95. We distinguish two cases:

(i) K has an archimedean absolute value. Then K has characteristic 0, by Corollary 1.1.3, and so K contains $\mathbf{Q}$ as prime subfield.

(ii) All the places on K are non-archimedean. Then the intersection $k = \cap \mathfrak{o}_\mathfrak{p}$, where $\mathfrak{p}$ runs over all places, is a subfield of K, for if $|\alpha|_\mathfrak{p} \leqslant 1$ for all $\mathfrak{p}$, this together with the product formula shows that $|\alpha|_\mathfrak{p} = 1$ for all $\mathfrak{p}$, hence $\alpha^{-1} \in k$, and this shows k to be a field. We shall call k the *field of constants*. All valuations are trivial on k and hence on its algebraic extensions; this shows K/k to be transcendental.

In case (i) we put $F = \mathbf{Q}$ and write ∞ for the absolute value. In case (ii) we put $F = k(x)$, where k is the field of constants and $x \in K$ is transcendental over k; here we shall use ∞ to indicate the place at ∞, corresponding to x^{-1}. We complete the proof by finding a bound on the degree $[K:F]$. Let $u_1, \ldots, u_n \in K$ be linearly independent over F and let $y \in F$ be any integer, i.e. rational integer in case (i), polynomial in x over k in case (ii). We define a family of real numbers $\alpha_\mathfrak{p}$ by the equations

$$\alpha_\mathfrak{p} = \begin{cases} \max_i \{|u_i|_\mathfrak{p}\} & \text{for } \mathfrak{p} \nmid \infty, \\ |y|_\infty^{e_\mathfrak{p}} (\Sigma_i |u_i|_\mathfrak{p}) & \text{for } \mathfrak{p} \mid \infty. \end{cases} \tag{15}$$

It is clear that $\alpha_\mathfrak{p} = 1$ for almost all $\mathfrak{p}$. Let $\mathfrak{A}$ be the parallelotope defined by $|\mathfrak{A}|_\mathfrak{p} = \alpha_\mathfrak{p}$. Then $L(\mathfrak{A})$ contains all elements $\Sigma a_i u_i$, where a_i is integral in F and $|a_i|_\infty \leqslant |y|_\infty$. For we have, by (15),

$$|\Sigma a_i u_i|_\mathfrak{p} \leqslant \begin{cases} \max_i \{|a_i|_\mathfrak{p} |u_i|_\mathfrak{p}\} \leqslant \alpha_\mathfrak{p} & \text{for } \mathfrak{p} \nmid \infty, \\ \max_i \{|a_i|_\mathfrak{p}\} (\Sigma |u_i|_\mathfrak{p}) \leqslant |y|_\mathfrak{p} (\Sigma |u_i|_\mathfrak{p}) \leqslant \alpha_\mathfrak{p} & \text{for } \mathfrak{p} \mid \infty. \end{cases}$$

Further, in case (ii) $L(\mathfrak{A})$ is clearly a k-space. For each a_i we have more than $|y|_\infty$ choices in each case (in case (ii), $|y|_\infty = q^\delta$ and a_i is a polynomial of degree at most δ). Hence, by Proposition 2.2,

$$|y|_\infty^n \leqslant \lambda(\mathfrak{A}) \leqslant \max\{1, M \,.\, \mu(\mathfrak{A})\}.$$

Now $\mu(\mathfrak{A}) = \Pi \alpha_\mathfrak{p} = C \,.\, |y|_\infty^r$, where $r = \sum_{\mathfrak{p} \mid \infty} e_\mathfrak{p}$ is a fixed integer. Hence for all sufficiently large $|y|_\infty$ we find

$$|y|_\infty^n < C \,.\, M \,.\, |y|_\infty^r.$$

It follows that $n \leqslant r$ and so $[K:F] \leqslant r$. Thus K is a finite extension of $\mathbf{Q}$ or $k(x)$, as claimed.■

We have now shown that every global field is a finite algebraic extension of **Q** or $k(x)$, and that every finite *separable* extension of **Q** or $k(x)$ is a global field. This still leaves the possibility of an inseparable extension in the function field case. Of course in characteristic 0 this cannot happen, and it remains to look at the case of finite characteristic. When the ground field is perfect, it is always possible to choose the transcendental element x in K such that K is a separable extension of $k(x)$. We recall this result and its proof (due to F. K. Schmidt):

Proposition 2.4
Let k be a perfect field and K an extension of k of transcendence degree 1. Then there exists $x \in K$ such that $K/k(x)$ is a separable extension.

An element x with the given property is called a *separating* element of the extension.

Proof
We note that k is invariantly defined in K as the field of constants, i.e. the intersection of all valuation rings, and that K is finitely generated over k of transcendence degree 1. Let $K = k(z_1, \ldots, z_r)$ and assume that k is perfect; we have to find $x \in K$ such that K is separable over $k(x)$. If $r = 1, K = k(z_1)$ is a rational function field and there is nothing to prove. Otherwise $r > 1$ and $z_1, \ldots, z_r$ are algebraically dependent. Take a polynomial relation of least degree

$$f(z_1, \ldots, z_r) = 0. \tag{16}$$

If some power of z_1 occurring in f is not divisible by p, then (16) is a separable polynomial for z_1 over $k(z_2, \ldots, z_r)$ and we can use induction on r to complete the proof. Similarly if some other z_i occurs to a power not divisible by p. This leaves only the case where

$$f(z_1, \ldots, z_r) = g(z_1^p, \ldots, z_r^p),$$

but then we can write $f = h^p$ for some polynomial h, because k is perfect and so all its elements are pth powers. Now h is a polynomial of lower degree and we can apply induction on the degree of f to complete the proof.■

In particular, this result covers the case of a finite field of constants, which apart from that of characteristic zero is the main case of interest. We shall slightly modify our definition of global field by excluding imperfect ground fields and defining a *global field* as a finite algebraic extension of either **Q** or of a rational function field $k(x)$, where k is a perfect field. By contrast, its completion at a place is sometimes called a *local field*.

Exercises

1. Let K be a global field with no archimedean places. Show that the field of constants is relatively algebraically closed in K.

2. Let K/k be a transcendental extension of degree 1 in characteristic p, where k is perfect (so that a separating element exists, by Proposition 2.4). Show that a separating element may be found by taking $\alpha \in K\backslash k$ and solving the equation $x^q = \alpha$ for the highest power $q = p^r$ of p.
3. Show that an element of a global field which is a uniformizer at a place must be a separating element.
4. Let $k = \mathbf{F}_p(s, t)$, where s, t are indeterminates and consider the extension E/k generated by x, y subject to the relation
$$y^p = x^{2p} + x^p s + t.$$
Show that there is no separating element for E over k.

3.3 THE UNIT THEOREM

In section 3.2 we obtained an upper bound for the number of points in a parallelotope in terms of its volume. In what follows we shall also need a lower bound for this number; in the simplest case such a lower bound may be a constant M such that any parallelotope of volume M contains at least two points. For convex regions symmetric about the origin (in algebraic number fields) this is of course just Minkowski's theorem, and the following result may be regarded as an analogue.

Theorem 3.1

Let K be a global field. Then there is a constant M such that for any parallelotope $\mathfrak{A}$ in K,

$$\mu(\mathfrak{A}) \leqslant M \,.\, \lambda(\mathfrak{A}), \tag{1}$$

where as before μ is the volume and λ the order.

Proof

As we have seen, K is a finite separable extension of a rational global field F ($= \mathbf{Q}$ or $k(x)$). We denote by A, B the rings of integers in F, K respectively. Here A is defined by absolute values in F, with an additional place at ∞ and B is the integral closure of A in K, which may also be defined by all the absolute values extending all the places other than ∞. By Theorem 2.6.2, B is free as A-module, with basis $u_1, \ldots, u_n$ say. We put

$$c_0 = \max \{|u_i|_{\mathfrak{p}}, \text{ where } \mathfrak{p} \mid \infty, i = 1, \ldots, n\}, \tag{2}$$

and in the function field case we denote the order of the residue class field at ∞ (essentially the constant field) by q, while for number fields $q = 2$. Given any parallelotope $\mathfrak{A}$, by the approximation theorem (Theorem 1.2.3) there exists $\alpha \in K$ such that

$$c_0 \leqslant |\alpha\mathfrak{A}|_{\mathfrak{p}} \leqslant qc_0 \quad \text{for all } \mathfrak{p} \mid \infty. \tag{3}$$

Now choose an integer y in $F^\times$ such that

$$|\alpha y\mathfrak{A}|_\mathfrak{p} \leqslant 1 \quad \text{for all } \mathfrak{p} \nmid \infty. \tag{4}$$

This is possible by the strong approximation theorem (Theorem 2.4.1). Since both $\lambda(\mathfrak{A})$ and $\mu(\mathfrak{A})$ are invariant under multiplication by a field element, we can replace $\mathfrak{A}$ by $\alpha y\mathfrak{A}$ in (3) and (4) and obtain

$$c_0|y|_\mathfrak{p} \leqslant |\mathfrak{A}|_\mathfrak{p} \leqslant qc_0|y|_\mathfrak{p} \quad \text{for } \mathfrak{p} \mid \infty, \tag{5}$$

$$|\mathfrak{A}|_\mathfrak{p} \leqslant 1 \quad \text{for } \mathfrak{p} \nmid \infty. \tag{6}$$

To find all the elements in $L(\mathfrak{A})$ (the 'lattice points'), we consider the set P of all elements

$$a_1u_1 + \ldots + a_nu_n \quad a_i \in A, |a_i|_\infty \leqslant |y|_\infty. \tag{7}$$

The possible number of choices for a_1 in the number field case is $2[|y|_\infty]+1$, while in the function field case the order is $q^{\deg y} = |y|_\infty$. Hence the number of elements in P (respectively the order of P) is at least $|y|_\infty^n$.

In what follows we repeatedly have to take products over certain sets of places; let us write Π' for products taken over all $\mathfrak{p} \nmid \infty$ and Π'' for products taken over all $\mathfrak{p} \mid \infty$, while (as before) Π stands for products taken over all $\mathfrak{p}$.

For any positive integers $n_\mathfrak{p}$ the additive group $B/\Pi'\mathfrak{p}^{n_\mathfrak{p}}$ has order $\Pi'(N\mathfrak{p})^{n_\mathfrak{p}}$. If we define $n_\mathfrak{p}$ by

$$|\mathfrak{A}|_\mathfrak{p} = (N\mathfrak{p})^{-n_\mathfrak{p}} \quad \text{for } \mathfrak{p} \nmid \infty, \tag{8}$$

then at least $|y|_\infty^n \Pi'(N\mathfrak{p})^{n_\mathfrak{p}}$ elements of P have the same image in $B/\Pi'\mathfrak{p}^{n_\mathfrak{p}}$, and for any $u, v \in P$ with the same image we have

$$|u-v|_\mathfrak{p} \leqslant |\mathfrak{A}|_\mathfrak{p} \quad \text{for } \mathfrak{p} \nmid \infty,$$

$$|u-v|_\mathfrak{p} \leqslant c_0|y|_\mathfrak{p} \leqslant |\mathfrak{A}|_\mathfrak{p} \quad \text{for } \mathfrak{p} \mid \infty,$$

by (7) and (5) respectively. Hence $u-v \in L(\mathfrak{A})$. If we fix $v \in P$ and let u run over P, we obtain $\lambda(\mathfrak{A}) \geqslant |y|_\infty^n / \Pi'(N\mathfrak{p})^{n_\mathfrak{p}}$, hence by (5) and (8),

$$\lambda(\mathfrak{A}) \geqslant |y|_\infty^n/\Pi'(N\mathfrak{p})^{n_\mathfrak{p}} \geqslant (qc_0)^{-n}\Pi''|\mathfrak{A}|_\mathfrak{p}/\Pi'(N\mathfrak{p})^{n_\mathfrak{p}} = (qc_0)^{-n}\mu(\mathfrak{A}).$$

Now (1) follows with $M = (qc_0)^n$; here c_0 is given by (2), while q is the order of the residue class field at ∞, or 2.■

For any parallelotope $\mathfrak{A}$ in a global field K with prescribed residue class field F we define the *absolute norm* of $\mathfrak{A}$ as

$$N\mathfrak{A} = \Pi'|\mathfrak{A}|_\mathfrak{p}^{-1}.$$

Thus if $|\mathfrak{A}|_\mathfrak{p} = (N\mathfrak{p})^{-n_\mathfrak{p}}$, then $N\mathfrak{A} = \Pi'(N\mathfrak{p})^{n_\mathfrak{p}}$, while for $\alpha \in K$ this gives $N(\alpha) = \Pi'|\alpha|_\mathfrak{p}^{-1} = \Pi''|\alpha|_\mathfrak{p}$, by the product formula. Hence for elements of K this agrees with the usual definition of the norm.

Our next aim is to prove the unit theorem, giving the structure of the group of units in a global field. We begin by taking all places, which gives us the *absolute units*, defined by

$$U_0 = \{\alpha \in K \,|\, |\alpha|_{\mathfrak{p}} = 1 \text{ for all } \mathfrak{p}\}.$$

Proposition 3.2
The group U_0 of absolute units in a global field with finite residue class fields is a finite cyclic group.

Here we have had to restrict ourselves to the case where $\lambda(\mathfrak{A})$ really means 'number of elements in $L(\mathfrak{A})$'. By Proposition 2.3 we are now dealing with finite extensions of $\mathbf{Q}$ or of $k(x)$, where k is a finite field; such a k is perfect and by taking x to be a separating element we can be sure (by Proposition 2.4) of our extension being separable.

Proof
It is clear that U_0 is contained in $L(\mathfrak{F})$, where $\mathfrak{F}$ is the parallelotope given by $|\mathfrak{F}|_{\mathfrak{p}} = 1$ for all $\mathfrak{p}$. More precisely, if $\alpha \in L(\mathfrak{F})$ and $\alpha \neq 0$, then $|\alpha|_{\mathfrak{p}} \leqslant 1$ for all $\mathfrak{p}$; since $\Pi |\alpha|_{\mathfrak{p}} = 1$, we must have $|\alpha|_{\mathfrak{p}} = 1$ for all $\mathfrak{p}$, and so $U_0 = L(\mathfrak{F}) \backslash \{0\}$.

By Proposition 2.2, $\lambda(\mathfrak{F})$ is finite, hence U_0 is finite, and as subgroup of a field must be cyclic, as we saw in section 3.1.■

In a global field K let us take any set S of places and define the group of S-units in K as

$$U(S) = \{\alpha \in K \,|\, |\alpha|_{\mathfrak{p}} = 1 \text{ for all } \mathfrak{p} \notin S\}.$$

Thus $U_0 = U(\varnothing)$ and the classical unit group is $U(S_\infty)$, where S_∞ denotes the set of all places at ∞, i.e. all $\mathfrak{p} \,|\, \infty$. We can now state the unit theorem in its general form:

Theorem 3.3 (*Unit theorem*)
Let K be a global field with finite residue class fields and let S be a finite set of places, including all places at ∞. If s denotes the number of places in S and U_0 is the absolute unit group, then the group of S-units is given by

$$U(S) \cong U_0 \times \mathbf{Z}^{s-1}.$$

Proof
Let $|\;|_i$ $(i = 1, \ldots, s)$ be the absolute values in S, write $v_i(\alpha) = -\log|\alpha|_i$ and consider the mapping

$$\log : U(S) \to \mathbf{R}^s, \qquad \alpha \mapsto (v_1(\alpha), \ldots, v_s(\alpha)).$$

Clearly this is a homomorphism with kernel U_0; so to complete the proof it will be enough to show that the image I is a free abelian group of rank $s - 1$. We note that the image lies in the hyperplane of $\mathbf{R}^s$ given by

$$\xi_1 + \ldots + \xi_s = 0, \tag{9}$$

by the product formula, so the result will follow if we can show that I is a discrete subgroup of $\mathbf{R}^s$ spanning the hyperplane (9). Here 'discrete' means a subgroup for which there is a neighbourhood of 0 containing no points of I apart from 0.

In the first place we note that by the upper estimate for $\lambda(\mathfrak{A})$ (Proposition 2.2) I has only finitely many elements in any bounded region of $\mathbf{R}^s$. We shall use induction on s. If $v_1(\alpha) = 0$ for all $\alpha \in U(S)$, then $I \subseteq \mathbf{R}^{s-1}$ and the result follows by induction. Otherwise take $\alpha_1 \in U(S)$ to be an element for which $v_1(\alpha)$ has the least positive value and write

$$I' = \{(v_2(\beta), \ldots, v_s(\beta)) \mid \beta \in U(S), v_1(\beta) = 0\}.$$

Then $I \cong \langle\alpha_1\rangle \times I'$, so by induction $I' \cong \mathbf{Z}^t$ with $t \leqslant s-2$, hence $I \cong \mathbf{Z}^r$, where $r \leqslant s-1$.

To complete the proof it will be enough to find $s-1$ linearly independent elements in $U(S)$. For this we shall need a lower estimate for $\lambda(\mathfrak{A})$, i.e. Theorem 3.1, but in a slightly sharper form, which we state as a lemma:

Lemma 3.4
Let K be a global field with finite residue class fields and let $\mathfrak{q}$ be a prescribed place of K. Then there exists a real positive constant $c = c(\mathfrak{q})$ such that for each parallelotope $\mathfrak{A}$ there is an $\alpha \in K$ satisfying

$$|\alpha|_{\mathfrak{p}} \leqslant |\mathfrak{A}|_{\mathfrak{p}} \leqslant c|\alpha|_{\mathfrak{p}} \quad \textit{for all } \mathfrak{p} \neq \mathfrak{q}. \tag{10}$$

Proof
By Theorem 3.1 we have $\lambda(\mathfrak{A}) > (2M)^{-1}\mu(\mathfrak{A})$. Let us take $\mathfrak{A}'$ to differ from $\mathfrak{A}$ only at $\mathfrak{q}$ and such that

$$1 \leqslant (2M)^{-1}\mu(\mathfrak{A}') \leqslant N\mathfrak{q} \quad (\text{where } N\mathfrak{q} = 1 \text{ if } \mathfrak{q} \text{ is archimedean}). \tag{11}$$

Then $\lambda(\mathfrak{A}') > 1$, so there exists $\alpha \neq 0$ in $L(\mathfrak{A}')$, and we have

$$|\alpha|_{\mathfrak{p}} \leqslant |\mathfrak{A}'|_{\mathfrak{p}} \quad \text{for all } \mathfrak{p}. \tag{12}$$

It follows that $|\alpha|_{\mathfrak{p}} \leqslant |\mathfrak{A}|_{\mathfrak{p}}$ for all $\mathfrak{p} \neq \mathfrak{q}$ and this proves the first inequality in (10). Further, by (12), $|\mathfrak{A}'\alpha^{-1}|_{\mathfrak{p}} \geqslant 1$ for all $\mathfrak{p}$, so if we fix $\mathfrak{p}$, then bearing in mind that $\mu(\mathfrak{A}'\alpha^{-1}) = \mu(\mathfrak{A}')$ by (6) of section 3.2, we have

$$|\mathfrak{A}'\alpha^{-1}|_{\mathfrak{p}} \leqslant \mu(\mathfrak{A}'\alpha^{-1}) = \mu(\mathfrak{A}') \leqslant 2M \,.\, N\mathfrak{q},$$

by (11). Hence for $\mathfrak{p} \neq \mathfrak{q}$ we have $|\alpha|_{\mathfrak{p}} \geqslant |\mathfrak{A}'|_{\mathfrak{p}}/2M \,.\, N\mathfrak{q}$, therefore (10) holds with $c = 2M \,.\, N\mathfrak{q}$, and the lemma follows.■

To continue with the proof of Theorem 3.3 we note that for any integer $c > 1$ there are only finitely many places $\mathfrak{p}$ satisfying $N\mathfrak{p} < c$. For if $\mathfrak{o}_S$ denotes the ring of S-integers (i.e. $v_{\mathfrak{p}}(\alpha) \geqslant 0$ for all $\mathfrak{p} \notin S$), take any $c+1$ integers

α_ν and suppose $\mathfrak{p}$ is such that $N\mathfrak{p} \leqslant c$. Then there must be two α's with the same image in $\mathfrak{o}_S/\mathfrak{p}$, thus $|\alpha_1 - \alpha_2|_\mathfrak{p} < 1$; but the number of places $\mathfrak{p}$ such that $|\alpha_i - \alpha_j|_\mathfrak{p} < 1$ for a fixed pair $i \neq j$ is finite, by the product formula.

When $S = \varnothing$, the result follows from Proposition 3.2, so we may take $s \geqslant 1$. We write $S = \{\mathfrak{p}_1, \ldots, \mathfrak{p}_s\}$ and apply the lemma with $\mathfrak{q} = \mathfrak{p}_1$. We can find a finite set T of non-archimedean places such that for $\mathfrak{p} \notin S \cup T$, $N\mathfrak{p} > c$, where c is chosen as in the lemma. For any place $\mathfrak{p}$ let $a_\mathfrak{p}$ be a value of $|\ |_\mathfrak{p}$; then the number of families $(a_\mathfrak{p})$ satisfying

$$c^{-1} \leqslant a_\mathfrak{p} \leqslant 1 \quad \text{for } \mathfrak{p} \in S \tag{13}$$

is finite. For when $\mathfrak{p} \notin T$, then $a_\mathfrak{p} = 1$ is the only possibility by the lemma, because $N\mathfrak{p} > c$, while for $\mathfrak{p} \in T$ there are only finitely many because the places in T have a discrete value group (the archimedean places are all in S). Let $\beta_1, \ldots, \beta_r \in K$ be chosen to realize all such possible distributions of values (using Theorem 1.2.3).

Next let $\mathfrak{A}$ be a parallelotope such that

$$|\mathfrak{A}|_{\mathfrak{p}_i} = \varepsilon \quad \text{for } i = 2, \ldots, s,$$

$$|\mathfrak{A}|_\mathfrak{p} = 1 \quad \text{for } \mathfrak{p} \in S,$$

where $\varepsilon > 0$ is to be fixed later. By Lemma 3.4 there exists $\alpha \in K$ such that

$$c^{-1}\varepsilon \leqslant |\alpha|_{\mathfrak{p}_i} \leqslant \varepsilon \quad \text{for } i = 2, \ldots, s, \tag{14}$$

$$c^{-1} \leqslant |\alpha|_\mathfrak{p} \leqslant 1 \quad \text{for } \mathfrak{p} \notin S. \tag{15}$$

By (15), α realizes one of the possible distributions (13), so for some j in the range $1 \leqslant j \leqslant r$ we have

$$|\alpha|_\mathfrak{p} = |\beta_j|_\mathfrak{p} \quad \text{for all } \mathfrak{p} \notin S.$$

Writing $\alpha = \beta_j u$, we have $|u|_\mathfrak{p} = 1$ for all $\mathfrak{p} \notin S$, so $u \in U(S)$, and by (14),

$$|u|_{\mathfrak{p}_i} = |\alpha|_{\mathfrak{p}_i}/|\beta_j|_{\mathfrak{p}_i} \leqslant \varepsilon/|\beta_j|_{\mathfrak{p}_i} \quad i = 2, \ldots, s. \tag{16}$$

Now the β's are fixed in advance and depend only on $\mathfrak{p}_1$, so we can choose ε so as to make the right-hand side of (16) small. Hence there exists $u \in U(S)$ such that $|u|_{\mathfrak{p}_i} < 1$ for $i = 2, \ldots, s$, and hence (by the product formula) $|u|_{\mathfrak{p}_1} > 1$.

Taking logarithms, we get a vector $e_1 \in \mathbf{R}^s$ with

$$v_1(e_1) > 0 > v_i(e_1) \quad (i \neq 1).$$

Similarly we find $e_j \in \mathbf{R}^s$ with

$$v_j(e_j) > 0 > v_i(e_j) \quad (i \neq j).$$

Writing these vectors as columns $y_1, \ldots, y_s$ say, $y_j = (y_{ij})$ with $\Sigma_i y_{ij} = 0$, we must show that there are $s-1$ linearly independent columns. Assume a relation between the last $s-1$ columns, say:

$$a_2 y_2 + \ldots + a_s y_s = 0, \text{ where } a_i \in \mathbf{R}.$$

By suitable numbering and changing sign if necessary we may assume that $a_s > 0$, $a_s \geqslant a_i$ $(i = 2, \ldots, s)$. Looking at the last row of (y_{ij}), we have, since $y_{is} < 0$ for $i \neq s$,

$$0 = a_2 y_{2s} + \ldots + a_s y_{ss} \geqslant a_s (y_{2s} + \ldots + y_{ss}) = -a_s y_{1s} > 0.$$

This contradiction shows that there are indeed $s-1$ linearly independent columns, as claimed.■

The unit theorem was first proved for number fields (with the set of all archimedean places for S) by Dirichlet in 1846; the proof was simplified by Minkowski and the general form (Theorem 3.3) is due to Chevalley [1940] and Hasse (1980).

Examples

(i) Let K be a finite extension of $\mathbf{Q}$ of degree n, and denote by r_1, r_2 the number of real and complex archimedean places, respectively. This means that there are r_1 homomorphisms of K into $\mathbf{R}$ and r_2 homomorphisms into $\mathbf{C}$, and we have $r_1 + 2r_2 = n$. Thus taking as S the set of all archimedean places, we have $r_1 + r_2 = s$ and

$$U(S) \cong U_0 \times \mathbf{Z}^{r_1 + r_2 - 1}.$$

For example, in a quadratic extension $(n = 2)$ $s = 2 - r_2 = 1$ or 2. Either $r_2 = 0$; then K is real and the group of units is the direct product of the (finite) group of roots of 1 in K times an infinite cyclic group. Or $r_2 = 1$; then K is complex and the group of units is finite.

(ii) Let K be a function field over a constant field k which is finite with q elements. The places at ∞ are the extensions of the degree valuation, $\mathfrak{p}_1, \ldots, \mathfrak{p}_s$ say, $\Sigma e_i f_i = n$. The unit group is the direct product of $k^\times$ and a free abelian group of rank $s-1$.

Exercises

1. Find the unit group in the field generated over $\mathbf{Q}$ by a primitive 12th root of 1 (cf. Exercise 5 of section 2.3).
2. Let p be an odd prime and write K for the field generated by a root $(\neq 1)$ of $x^p = 1$. Show that the degree of K is $p-1$, the number of absolute values on K extending ∞ is $s = (p+1)/2$ and the discriminant is (p^{p-2}).
3. Find the unit group in the field generated over a field k of characteristic $\neq 2$ by x, y subject to $x^2 + y^2 = 1$.
4. Let d be a positive square free integer. Show that the solutions of

$x^2 - dy^2 = 1$ in integers form an infinite cyclic group (this is known as Pell's equation).

3.4 THE CLASS NUMBER

We have seen that every global field K is a finite algebraic extension (separable by hypothesis) of a rational global field F, where $F = \mathbf{Q}$ or $k(x)$. The ring A of integers in F is obtained by omitting one place (the place 'at ∞') and in K we have the integral closure B of A in K. If $\mathfrak{p}$ is a place on F and $\mathfrak{P}$ a place on K dividing $\mathfrak{p}$, then we have seen that

$$N_{K/F}(\mathfrak{P}) = \mathfrak{p}^{f_{\mathfrak{p}}}. \tag{1}$$

The absolute norm of $\mathfrak{P}$ is defined as

$$N\mathfrak{P} = (N\mathfrak{p})^{f_{\mathfrak{p}}}, \tag{2}$$

where $N\mathfrak{p}$ is the order of $A/\mathfrak{p}$. Since $f_{\mathfrak{p}}$ is the residue degree of $\mathfrak{P}$, we see that $N\mathfrak{P}$ is just the order of the residue class field $B/\mathfrak{P}$ (i.e. either the number of elements of $B/\mathfrak{P}$ or in the case of a function field with infinite constant field, q^r, where q is the order of k and r the degree over k). For any fractional ideal $\mathfrak{A}$ in B we define the absolute norm by multiplicativity in terms of equation (2):

$$\text{If } \mathfrak{A} = \mathfrak{P}_1^{\alpha_1} \ldots \mathfrak{P}_t^{\alpha_t}, \quad \text{then} \quad N\mathfrak{A} = N\mathfrak{P}_1^{\alpha_1} \ldots N\mathfrak{P}_t^{\alpha_t}.$$

In terms of the $\mathfrak{P}$-adic absolute value this can be written as

$$N\mathfrak{A} = \left(\prod_{\mathfrak{P} \nmid \infty} |\mathfrak{A}|_{\mathfrak{P}} \right)^{-1}. \tag{3}$$

We note in particular:

1. If $\mathfrak{A}$ is an integral ideal, then the above derivation shows that $N\mathfrak{A}$ is the order of $B/\mathfrak{A}$;
2. If $\mathfrak{A}$ is a principal ideal, $\mathfrak{A} = (\alpha)$, then by (3) and the product formula,

$$N(\alpha) = \prod_{\mathfrak{P} \mid \infty} |\alpha|_{\mathfrak{P}}. \tag{4}$$

In other words, $N(\alpha)$ is the product of the absolute values of α.

We recall (Theorem 2.5.1) that B is a Dedekind domain, so the fractional ideals form a group D under multiplication, with the subgroup D_0 of principal ideals. The quotient $C = D/D_0$ is called the *ideal class group*. Our main result in this section states that C is finite; its order is the *class number* of K.

Theorem 4.1
The ideal class group of any global field is finite.

Proof

The lower estimate for the order of a parallelotope $\mathfrak{A}$ in K (Theorem 3.3) shows that

$$\lambda(\mathfrak{A}) \geqslant M^{-1}\mu(\mathfrak{A}). \tag{5}$$

We take a finite set S of places of K, including all the archimedean ones, and use Π'' for the product over the places in S and Π' for the product over the rest. Fix $\mathfrak{q} \in S$ and for any fractional ideal $\mathfrak{A}$ in the ring $\mathfrak{o}_S$ of S-integers define a parallelotope $\tilde{\mathfrak{A}}$ by the equations

$$|\tilde{\mathfrak{A}}|_{\mathfrak{p}} = |\mathfrak{A}|_{\mathfrak{p}} \quad \text{for } \mathfrak{p} \notin S,$$

while for $\mathfrak{p} \in S$, $|\tilde{\mathfrak{A}}|_{\mathfrak{p}}$ is adjusted so that

$$1 \leqslant (2M)^{-1}\mu(\tilde{\mathfrak{A}}) \leqslant N\mathfrak{q}, \tag{6}$$

where $N\mathfrak{q} = 1$ if $\mathfrak{q}$ is archimedean. This is always possible because $|\tilde{\mathfrak{A}}|_{\mathfrak{q}}$ varies by powers of $N\mathfrak{q}$. By (5) and (6) we have

$$\lambda(\tilde{\mathfrak{A}}) > (2M)^{-1}\mu(\tilde{\mathfrak{A}}) \geqslant 1,$$

so there exists $\alpha \in \mathfrak{A}$, $\alpha \neq 0$, such that

$$\begin{aligned} N(\alpha) &= \prod{}'' |\alpha|_{\mathfrak{P}} \\ &\leqslant \prod{}'' |\mathfrak{A}|_{\mathfrak{P}} \\ &= \mu(\tilde{\mathfrak{A}})/\prod{}' |\tilde{\mathfrak{A}}|_{\mathfrak{P}} \\ &= \mu(\tilde{\mathfrak{A}}) \,.\, N(\mathfrak{A}) \\ &\leqslant (2M \,.\, N\mathfrak{q})N(\mathfrak{A}), \end{aligned}$$

by (6). Thus any fractional ideal $\mathfrak{A}$ contains an element α such that

$$0 < N(\alpha) < c \,.\, N(\mathfrak{A}), \tag{7}$$

where c is fixed and depends only on S. If we apply the argument to $\mathfrak{A}^{-1}$ we obtain $\beta \in \mathfrak{A}^{-1}$ such that $0 < N(\beta) < c \,.\, N(\mathfrak{A}^{-1})$, hence $\beta\mathfrak{A} \subseteq B$ and $N(\beta\mathfrak{A}) \leqslant c$. In other words, in every ideal class there is an integral ideal with norm $\leqslant c$. Now the number of such ideals is finite; for by the remark after Lemma 3.4, the number of prime ideals with this property is finite, and only finitely many power products are allowed. Hence the number of ideal classes is also finite.■

By a more precise calculation it can be shown that the constant c in (7) may be taken to be

$$c = \frac{n!}{n^n}\left(\frac{4}{\pi}\right)^t D^{1/2}, \tag{8}$$

where D is the discriminant of the field and t the number of complex absolute values (i.e. complex embeddings of K). The factor multiplying $D^{1/2}$ in (8) is known as *Minkowski's constant* (cf. Lang (1970)).

Since $\alpha \in \mathfrak{A}$ in (7), we have $(\alpha) \subseteq \mathfrak{A}$ and so $(\alpha) = \mathfrak{A}\mathfrak{B}$ for a fractional ideal $\mathfrak{B}$. It follows that $N(\mathfrak{B}) < c$, but we also have $N(\mathfrak{B}) \geqslant 1$ and so

$$D \geqslant \frac{n^{2n}}{n!^2}\left(\frac{\pi}{4}\right)^{2t}.$$

Here the right-hand side is easily seen to be > 1 whenever $n > 1$. This shows that every algebraic number field is ramified over **Q**. For relative extensions of number fields this need no longer hold, thus there are extensions L/K whose relative discriminant is a unit. For example, the splitting field K of $x^5 - x + 1 = 0$ over **Q** has discriminant $D = 2869 = 19.151$ and is unramified over $\mathbf{Q}(\sqrt{D})$ (cf. Lang (1970)).

The use of Minkowski's constant also leads to more precise estimates for the class number; these can still be improved with the help of the Dedekind zeta-function (cf. Borevich and Shafarevich (1966)).

In Theorem 4.1, let us take a transversal for the subgroup D_0 in D, say $\mathfrak{A}_1, \ldots, \mathfrak{A}_r$ and adjoin to S all places occurring with non-zero exponent in one of the $\mathfrak{A}_i$; then all classes become principal. We thus find:

Corollary 4.2

Let K be a global field. Given any finite set S of places on K, there is a finite set $S_1 \supseteq S$ such that for any set of places including S_1 the class number is 1. Hence the ring of all S_1-integers is a principal ideal domain.■

Exercises

1. In $\mathbf{Q}(\sqrt{-5})$ show that $(2, 1+\sqrt{-5})^2 = (2)$; verify more generally that the class number is two.
2. Let m, n be two squarefree numbers that are coprime. Show that in the ring of all integers of $\mathbf{Q}(\sqrt{m}, \sqrt{n})$ the ideal $(\sqrt{m}, \sqrt{n})$ is the unit ideal.
3. Let K be an algebraic number field with class number $h > 1$. Show that for any fractional ideal $\mathfrak{A}$ in K, $\mathfrak{A}^r$ is principal for some integer r dividing h. If $\mathfrak{A}^r = (\alpha)$, show that $\mathfrak{A}$ becomes principal in $K(\alpha^{1/r})$. Deduce that there is a finite extension L of K such that every fractional ideal of K becomes principal in L. (In general the ideals in L need not all be principal. Moreover, Golod and Shafarevich in 1964 constructed infinite 'class field towers' $K_1 \subset K_2 \subset \ldots$, i.e. algebraic number fields such that every ideal of K_r becomes principal in K_{r+1} but there is no algebraic number field with class number 1 to contain K_1. This answers a problem originally posed by Hilbert.)

4

Function fields

In Chapter 3 we have seen that an algebraic function field of one variable is closely analogous to an algebraic number field, though there are also some number-theoretic methods available for the latter which have no parallel for the former. Instead there are two other sources we can call on in the study of function fields: complex function theory and algebraic geometry. A full development of the geometric connexion would go beyond the framework of this book, so apart from occasional geometric indications we shall concentrate on the function-theoretic approach.

Each algebraic function field in one variable may (with **C** as ground field) be represented by a Riemann surface or, from the geometric point of view, as the field of rational functions on a certain algebraic curve. The places of our function field over the constant field correspond to the points of the Riemann surface. Our main objective, a description of the divisor class group, is accomplished by the Abel–Jacobi theorem. This is a result of function theory, though the actual proofs will be algebraic, as far as possible.

4.1 DIVISORS ON A FUNCTION FIELD

Let k be any field. An *algebraic function field* over k is a finitely generated extension field K of k which is not algebraic. If the transcendence degree is r, tr. deg. $K/k = r$, then K is a *field of r variables* over k. In what follows we shall take $r = 1$ and assume that k is perfect; as we have seen in section 3.2, there is then a separating element in K, i.e. an element $x \in K\backslash k$ such that K is separable over $k(x)$. Thus for the rest of this chapter, *function field* will mean 'algebraic function field of one variable over a perfect ground field'. As a finite separable extension $K/k(x)$ can be generated by a single element y; the minimal equation for y over $k(x)$ defines an algebraic curve, and in this sense a function field represents the field of all rational functions on an algebraic curve. For example, the quadratic extension of $k(x)$ generated by

$y = \sqrt{(1 - x^2)}$ is the function field of the circle $x^2 + y^2 = 1$. Most of the subsequent results can be stated either for function fields or for algebraic curves.

We recall that the field of constants in K is the subfield k' on which all the valuations of K over k are trivial; thus $k' \supseteq k$. This field k' can also be characterized as the relative algebraic closure of k in K. For simplicity we shall usually assume that $k' = k$; this makes it unnecessary to refer to the ground field explicitly. This condition will hold automatically whenever k is algebraically closed, or even just (perfect and) relatively algebraically closed in K.

Let K/k be a function field. We recall that a valuation on K is *over* k if its restriction to k is trivial, and that a *place* on K/k is an equivalence class of valuations on K/k. We note that if $x \in K \backslash k$, then:

1. All valuations of $k(x)/k$ are principal (i.e. discrete of rank 1, cf. Proposition 1.4.4);
2. All valuations of K/k are obtained by extending those of $k(x)/k$ (Theorem 2.3.3). They are again principal, and there are infinitely many (cf. Proposition 1.4 below), finitely many for each valuation of $k(x)/k$. Although the latter depend on x, the set of all valuations of K/k does not depend on the choice of x.

Let $\mathfrak{p}$ be any place of K/k, with associated valuation v and residue class field $k_{\mathfrak{p}}$. Then $k_{\mathfrak{p}}$ is a finite algebraic extension of k; for we have $v(\alpha) = 0$ for all $\alpha \in k^{\times}$, so the residue class map $a \mapsto \bar{a}$ of K into $k_{\mathfrak{p}} \cup \{\infty\}$ is injective when restricted to k, hence k is embedded in $k_{\mathfrak{p}}$. The restriction to $k(x)$ has as residue class field a finite extension of k, of degree equal to the degree of the irreducible polynomial defining the place p on $k(x)$ such that $\mathfrak{p} \mid p$. It follows that $f_{\mathfrak{p}} = [k_{\mathfrak{p}} : k]$ is finite too; we shall call it the *absolute residue degree* or simply the *degree* of $\mathfrak{p}$. Its relation to the relative degree is given by the formula

$$[k_{\mathfrak{p}} : k] = [k_{\mathfrak{p}} : k_p][k_p : k], \qquad (1)$$

where $\mathfrak{p} \mid p$. We shall also say that $\mathfrak{p}$ lies *over* p, a usage that will be justified when we come to the geometric interpretation. Of course when k is algebraically closed, we have $f_{\mathfrak{p}} = 1$.

As an example let p be the place defined by the linear polynomial $x - \alpha$, where $\alpha \in k$. At the place p the element $u = u(x)$ of $k(x)$ takes the value $u(\alpha)$. In particular, the valuation ring of p consists of those u that are finite at $x = \alpha$, while the maximal ideal consists of all u vanishing for $x = \alpha$. Similarly at the place corresponding to x^{-1}, the valuation ring consists of those u that are finite for $x = \infty$ while the maximal ideal consists of all u vanishing at ∞.

The field K is generated by an element y over $k(x)$ (by the theorem of the primitive element, Theorem 3.9.2 in A.2 which applies because $K/k(x)$ is separable). Thus we have $K = k(x, y)$, where x and y are related by an equation

$$f(x, y) = y^n + a_1 y^{n-1} + \ldots + a_n = 0, \quad a_i \in k(x). \qquad (2)$$

In effect f is the minimal polynomial for y over $k(x)$. Let δ be the discriminant of f, *qua* polynomial in y; thus δ is a rational function of x, whose zeros are the values of x at which (2) has multiple roots. These values are called the *branch points* or *ramification points* of y, the other points being *regular*. Since y is separable over $k(x)$, its minimal polynomial has no repeated factors, so δ does not vanish identically and it follows that y has only finitely many branch points. For example, $x^2 + y^2 = 1$ has the discriminant $4(1 - x^2)$ and so the branch points are at $x = \pm 1$; this shows that the branch points in general depend on the choice of x.

At any point $x = \alpha$ at which $f(\alpha, y)$ has distinct zeros $\beta_1, \ldots, \beta_n$, we can expand the solutions of (2) as power series

$$y^{(i)} = \beta_i + \sum c_{i\nu}(x - \alpha)^{\nu}.$$

At a branch point this is replaced by an expansion in fractional powers of $x - \alpha$; if the ramification index is r and t is a uniformizer, then $x - \alpha = t^r$ and we have a power series in t. At a point where one or more of the coefficients of f in (2) has a pole, we obtain for y an expansion involving some negative powers of x, or if we are at a branch point, negative fractional powers of x. These expansions play an important role in the function-theoretic approach, as well as in the closer geometric study of the algebraic curve defined by (2).

Given any $u \in K$ and any place $\mathfrak{p}$ of K/k, with normalized valuation $v_{\mathfrak{p}}$, there are three possibilities:

(i) If $v_{\mathfrak{p}}(u) > 0$, we say that u has a *zero* at $\mathfrak{p}$ of order $v_{\mathfrak{p}}(u)$;

(ii) If $v_{\mathfrak{p}}(u) < 0$, we say that u has a *pole* at $\mathfrak{p}$ of order $-v_{\mathfrak{p}}(u)$;

(iii) $v_{\mathfrak{p}}(u) = 0$; in this case, writing $a \mapsto \bar{a}$ for the residue class map, we obtain $\bar{u} \in k_{\mathfrak{p}}$, and this is called the *value* of u at the place $\mathfrak{p}$.

When $k = \mathbf{C}$, the complex field, the places of $\mathbf{C}(x)/\mathbf{C}$ are all the elements of $\mathbf{C}$ together with ∞, making up the Riemann sphere. A function field $K = \mathbf{C}(x, y)$ is a covering surface of the sphere, a *Riemann surface*, which is an n-fold covering, where $n = [K : \mathbf{C}(x)]$; this means that above each point of the Riemann sphere there lie n points of the Riemann surface (counted with their multiplicities, given by the ramification index). This is expressed by the conorm map (cf. (8) of section 2.5):

$$p \mapsto \prod_{\mathfrak{p} \mid p} \mathfrak{p}^{e_{\mathfrak{p}}}. \tag{3}$$

We also recall the norm map from section 2.5:

$$N(\mathfrak{p}) = p^{f_{\mathfrak{p}}}, \tag{4}$$

bearing in mind that $f_{\mathfrak{p}} = 1$ whenever the ground field is algebraically closed.

Let us recall the approximation theorem (Theorem 1.2.3); in the context of function fields this can be stated as follows: given any distinct elements $a_1, \ldots, a_m$ of K and any distinct places $\mathfrak{p}_1, \ldots, \mathfrak{p}_m$, there exists $u \in K$ which approximates a_i at $\mathfrak{p}_i$ for $i = 1, \ldots, m$ to any prescribed degree of accuracy; thus given $N > 0$, we can find u such that $u - a_i$ has a zero of order at least N at $\mathfrak{p}_i$, for $i = 1, \ldots, m$. If we take all the places of K which do not lie over ∞, we have a Dedekind ring, and by the strong approximation theorem we can then choose the function u approximating a_i at $\mathfrak{p}_i$ as before and also to be integral (i.e. remain finite) at every $\mathfrak{p} \neq \mathfrak{p}_i$. For $k = \mathbf{C}$ this shows: given a finite set of points $\mathfrak{p}_1, \ldots, \mathfrak{p}_m$ in the complex plane there exists a meromorphic function with assigned principal parts at these points and otherwise holomorphic in the finite plane (Mittag–Leffler's theorem). This freedom to manœuvre is of course due to the fact that we have not prescribed the behaviour at ∞. Once we include ∞ we find that the analogue no longer holds because for any function the sum of all its orders must vanish: $\Sigma\, v_{\mathfrak{p}}(u) = 0$. This is of course nothing else but the product formula, now written additively. When this condition is satisfied we can always construct a function in the rational case, $K = k(x)$. In the general case of function fields of one variable this is no longer so; the conditions for a function with prescribed poles and zeros to exist are given by the Riemann–Roch theorem (cf. section 4.6). It may nevertheless be possible to construct such functions by taking suitable transcendental extensions, but this will not concern us here (cf. Hensel and Landsberg, (1902)).

Let k be an algebraically closed field and consider a function field K over k. We take a separating element x of K and write $E = k(x)$; then any element y of K is separable over E. Consider the completion $\hat{E} = k((x))$ at $x = 0$; by section 2.3 we have

$$K \otimes \hat{E} \cong K_1 \times \ldots \times K_r,$$

where each K_i is a finite extension of $\hat{E}$, of degree e_i (by Theorem 2.3.3, recall that now $f_i = 1$). Thus K_i consists of all Laurent series in x^{1/e_i}. We shall assume that e_i is prime to the characteristic of k, to ensure that x^{1/e_i} is separable over E (this case is often described as *tamely ramified*). We thus obtain the following precise version of Puiseux's theorem:

Theorem 1.1
Let k be an algebraically closed field and x an indeterminate. Given y, separably algebraic over $k(x)$, suppose that the ramification index e at a place of $k(x, y)$ over the place $x = 0$ of $k(x)$ is not divisible by char k. Then y can be written as a power series in $x^{1/e}$.

In characteristic 0 this reduces to the usual form of Puiseux's theorem (van der Waerden (1939)):

Corollary 1.2
Let k be an algebraically closed field of characteristic 0, and with an

indeterminate x, write $K_r = k((x^{1/r}))$. Then the union $\cup K_r$ is an algebraically closed field.

The theorem is clear, from the remarks made earlier. To obtain the corollary, write $L = \cup K_r$. Since the K_r form a directed system: $K_r \cup K_s \subseteq K_{rs}$, it is clear that L is a field. Any element u algebraic over L is algebraic over some K_r. Writing $y = x^{1/r}$, we have an algebraic extension of $k((y))$, and by Theorem 1.1, this is generated by $y^{1/s}$, for some s, hence u lies in K_{rs}.■

For function fields the estimate for the different obtained in Theorem 2.6.6 can also be made more precise:

Proposition 1.3
Let K/k be a function field, where k is algebraically closed. Given a separating element x of K, let $\mathfrak{p}$ be any place of K, with uniformizer t and ramification index e of K over $k(x)$ at $\mathfrak{p}$. If e is not divisible by the characteristic of k, then the different $\mathfrak{d} = \mathfrak{d}_{K/k(x)}$ satisfies

$$(\mathfrak{d})_{\mathfrak{p}} = (t^{e-1}). \tag{5}$$

Proof
Let $\mathfrak{p}$ lie over the place p of $k(x)$, write A, B for the valuation rings of p, $\mathfrak{p}$ and take a uniformizer z for p. We assert that the complement B' of B satisfies $B' = Bt^{1-e}$; the result will then follow by the formula $\mathfrak{d} = (B')^{-1}$, cf. (7) of section 2.6.

The elements of B are power series in $z^{1/e}$, so we have an expansion

$$t^{1-e} = \Sigma b_\nu z^{\nu/e}, \quad b_{1-e} \neq 0. \tag{6}$$

Now for any $u \in B$, $\mathrm{Tr}(ut^{1-e})$ is a Laurent series in z; the lowest term in t is

$$z^{(1-e)/e} = z^{(1/e)-1},$$

while u has only non-negative powers, hence $Tr(ut^{1-e})$ is an integral power series in z and so lies in A. Hence $Bt^{1-e} \subseteq B'$; if this inequality is strict, suppose that $Bt^{-m} \subseteq B'$ with $m \geqslant e$. Take $a \in B$ of order $m-e$ and expand at^{-m} in powers of $z^{1/e}$; we find

$$at^{-m} = c_0 z^{-1} + \ldots, \quad c_0 \neq 0,$$

and since char $k \nmid e$,

$$\mathrm{Tr}(at^{-m}) = ec_0 z^{-1} + \ldots, \quad \notin A.$$

Thus $at^{-m} \notin B'$, and so $Bt^{1-e} = B'$, as we wished to show.■

The places where $e > 1$ are just the branch points or ramification points, and $\mathfrak{d}$ is also called the *ramification divisor*.

Let us now examine the set of all places on a function field more closely. We begin by showing that every function field has infinitely many places:

Proposition 1.4
Every function field has infinitely many places.

Proof
If the constant field k is infinite, then $k(x)$ has infinitely many places of degree 1. If k is finite, then there are infinitely many irreducible polynomials over k, by a well-known argument going back to Euclid: if there were only finitely many irreducible polynomials over k, say $f_1, \ldots, f_r$, then $f_1 \ldots f_r + 1$ would be a polynomial of positive degree and not divisible by any f_i, and so would contain another irreducible polynomial as factor, a contradiction.■

It is also easy to derive an analogue of Liouville's theorem:

Proposition 1.5
In a function field, any non-constant element has finitely many zeros and at least one zero; similarly for poles.

Proof
Given $x \in K \backslash k$, x defines a place on $k(x)$ which has finitely many extensions to K, and each of these is a zero of x. Taking x^{-1}, we see that the same holds for poles.■

We remark that Proposition 1.5 shows that any two functions agreeing at infinitely many places coincide, since their difference has infinitely many zeros. Likewise two functions coincide if they have the same expansion at one place, for their difference has a zero of infinite order at that place and so must vanish.

In what follows we shall often want to talk of the set of all functions having zeros of order at least ν_i (or poles of order at most $-\nu_i$) at given places $\mathfrak{p}_i$ $(i = 1, \ldots, r)$. For this purpose it is convenient to use the divisors introduced in section 2.4. We recall that the *divisor group D* is the free abelian group on the set of all places as free generating set; its elements are called *divisors*. A place is also called a *prime divisor* and the general divisor has the form

$$\mathfrak{a} = \mathfrak{p}_1^{\alpha_1} \ldots \mathfrak{p}_r^{\alpha_r} = \prod \mathfrak{p}^{v_{\mathfrak{p}}(\mathfrak{a})}. \tag{7}$$

If $\mathfrak{a}$ is given by (7), we also say: $\mathfrak{a}$ is *based* on the set $\{\mathfrak{p}_1, \ldots, \mathfrak{p}_r\}$. Geometrically a divisor represents a finite system of points (with given multiplicities) on the curve defined by the function field.

To express the fact that $u \in K$ has a zero of order at least α_i at $\mathfrak{p}_i$ $(i = 1, \ldots, r)$ we shall write

$$u \equiv 0 \pmod{\mathfrak{a}}, \tag{8}$$

where $\mathfrak{a}$ is given by (7).

We shall write $\mathfrak{e}$ for the *unit divisor*, defined by the equations $v_{\mathfrak{p}}(\mathfrak{e}) = 0$ for all $\mathfrak{p}$. If in (7), $\alpha_i \geqslant 0$ for all i, $\mathfrak{a}$ is called *integral*, and we say that $\mathfrak{a}$ *divides* $\mathfrak{b}$ and write $\mathfrak{a}|\mathfrak{b}$ if $\mathfrak{b}\mathfrak{a}^{-1}$ is integral. Two divisors $\mathfrak{a}$, $\mathfrak{b}$ are called *coprime* if $v_\mathfrak{p}(\mathfrak{a})v_\mathfrak{p}(\mathfrak{b}) = 0$ for all $\mathfrak{p}$; we note that this definition applies even if $\mathfrak{a}$, $\mathfrak{b}$ are not integral. Any divisor can be written as a quotient of two integral divisors that are coprime: $\mathfrak{a} = \mathfrak{n}/\mathfrak{d}$; here $\mathfrak{n}$ is called the *numerator* and $\mathfrak{d}$ the *denominator* of $\mathfrak{a}$. Further, we define the highest common factor (HCF) and the least common multiple (LCM) of $\mathfrak{a}$ and $\mathfrak{b}$ by

$$\mathrm{HCF}(\mathfrak{a}, \mathfrak{b}) = \prod \mathfrak{p}^{\lambda_\mathfrak{p}}, \quad \text{where } \lambda_\mathfrak{p} = \min\{v_\mathfrak{p}(\mathfrak{a}), v_\mathfrak{p}(\mathfrak{b})\},$$

$$\mathrm{LCM}(\mathfrak{a}, \mathfrak{b}) = \prod \mathfrak{p}^{\mu_\mathfrak{p}}, \quad \text{where } \mu_\mathfrak{p} = \max\{v_\mathfrak{p}(\mathfrak{a}), v_\mathfrak{p}(\mathfrak{b})\}.$$

Each divisor has a *degree*, given by

$$d(\mathfrak{a}) = \sum f_\mathfrak{p} v_\mathfrak{p}(\mathfrak{a}),$$

where $f_\mathfrak{p}$ is the degree of $\mathfrak{p}$ as defined earlier. Of course when k is algebraically closed, then $f_\mathfrak{p} = 1$ and it can be omitted. We note that the mapping $\mathfrak{a} \mapsto d(\mathfrak{a})$ defines a homomorphism from the divisor group to $\mathbf{Z}$; the kernel is the group D_0 of divisors of degree zero.

For each divisor $\mathfrak{a}$ we define a vector space over k by the equation

$$L(\mathfrak{a}) = \{u \in K \mid u \equiv 0 \pmod{\mathfrak{a}}\}. \tag{9}$$

If $\mathfrak{a}$ is an integral divisor other than the unit divisor, then $L(\mathfrak{a}) = 0$; for $u \in L(\mathfrak{a})$ means that $v_\mathfrak{p}(u) \geqslant v_\mathfrak{p}(\mathfrak{a}) \geqslant 0$, with strict inequality for at least one $\mathfrak{p}$, hence by the product formula, u must then be 0. On the other hand, to say that $u \in L(\mathfrak{a}^{-1})$ means that the poles of u are limited by the conditions $v_\mathfrak{p}(u) \geqslant -v_\mathfrak{p}(\mathfrak{a})$, or more briefly, the poles of u are confined to $\mathfrak{a}$.

The spaces (9), and more particularly $L(\mathfrak{a}^{-1})$ for an integral divisor $\mathfrak{a}$, are of great importance in what follows; to facilitate their study we shall need other related spaces. Given any set S of places, we put

$$L(\mathfrak{a} \mid S) = \{u \in K \mid v_\mathfrak{p}(u) \geqslant v_\mathfrak{p}(\mathfrak{a}) \text{ for all } \mathfrak{p} \in S\}.$$

The defining condition can also be expressed by writing $u \equiv 0 \pmod{\mathfrak{a} \mid S}$. These spaces $L(\mathfrak{a} \mid S)$ are generally larger than $L(\mathfrak{a})$ (because they are defined by fewer conditions); this makes them easier to handle. We note the following obvious properties:

1. $L(\mathfrak{a}) = L(\mathfrak{a} \mid \Sigma)$, *where* Σ *is the set of all places;*
2. $S \subseteq S' \Rightarrow L(\mathfrak{a} \mid S) \supseteq L(\mathfrak{a} \mid S')$;
3. $\mathfrak{a} \mid \mathfrak{b} \Leftrightarrow v_\mathfrak{p}(\mathfrak{a}) \leqslant v_\mathfrak{p}(\mathfrak{b})$ *all* $\mathfrak{p} \Rightarrow L(\mathfrak{a} \mid S) \supseteq L(\mathfrak{b} \mid S)$;
4. *If* $\mathfrak{a}\mathfrak{b}^{-1}$ *is based outside* S, *then* $L(\mathfrak{a} \mid S) = L(\mathfrak{b} \mid S)$.

Our first aim is to show that $L(\mathfrak{a})$ is finite-dimensional over k; to achieve it we begin by computing some relative dimensions.

Lemma 1.6

Given a finite set S of places on a function field K and divisors $\mathfrak{a}$, $\mathfrak{b}$ such that $\mathfrak{a}|\mathfrak{b}$, denote by $\mathfrak{a}'$, $\mathfrak{b}'$ the divisors obtained from $\mathfrak{a}$, $\mathfrak{b}$ by omitting all places outside S. Then

$$\dim_k(L(\mathfrak{a} \mid S)/L(\mathfrak{b} \mid S)) = d(\mathfrak{b}'\mathfrak{a}'^{-1}) = d(\mathfrak{b}') - d(\mathfrak{a}'). \tag{10}$$

Proof

By Remark 4 above we may replace $\mathfrak{a}$, $\mathfrak{b}$ by $\mathfrak{a}'$, $\mathfrak{b}'$ and then omit the primes. We now prove the result by induction on the number of factors in $\mathfrak{b}\mathfrak{a}^{-1}$. Suppose first that $\mathfrak{b} = \mathfrak{a}\mathfrak{q}$, where $\mathfrak{q} \in S$; we have to show

$$\dim_k(L(\mathfrak{a} \mid S)/L(\mathfrak{a}\mathfrak{q} \mid S)) = f_{\mathfrak{q}}. \tag{11}$$

Write $f = f_{\mathfrak{q}}$ and take $x_1, \dots, x_f \in \mathfrak{o}_{\mathfrak{q}}$ such that the residues $\bar{x}_1, \dots, \bar{x}_f$ in the residue class field $k_{\mathfrak{q}}$ form a k-basis. By the approximation theorem we can find $x'_1, \dots, x'_f \in K$ such that

$$v_{\mathfrak{q}}(x_i - x_i') > 0, \qquad v_{\mathfrak{p}}(x_i') \geqslant 0 \quad \text{for all } \mathfrak{p} \in S, \mathfrak{p} \neq \mathfrak{q}.$$

It follows that $x'_1, \dots, x'_f \in \mathfrak{o}_{\mathfrak{q}}$ and x'_i has the same image as x_i in $k_{\mathfrak{q}}$. We can therefore replace x_i by x'_i and then drop the primes; thus we may assume henceforth that $x_i \in \mathfrak{o}_{\mathfrak{p}}$ for all $\mathfrak{p} \in S$.

Secondly, again by the approximation theorem, there exists $u \in K$ such that $v_{\mathfrak{p}}(u) = v_{\mathfrak{p}}(\mathfrak{a})$ for all $\mathfrak{p} \in S$, because S is finite. It follows that

$$ux_1, \dots, ux_f \in L(\mathfrak{a} \mid S);$$

we claim that the ux_i form a basis of $L(\mathfrak{a} \mid S)$ mod $L(\mathfrak{a}\mathfrak{q} \mid S)$.

Let $y \in L(\mathfrak{a} \mid S)$; then $u^{-1}y$ is integral at all $\mathfrak{p} \in S$, hence there exist $a_1, \dots, a_f \in k$ such that $v_{\mathfrak{q}}(u^{-1}y - \Sigma a_i x_i) > 0$ and so $y - \Sigma a_i u x_i \in L(\mathfrak{a}\mathfrak{q} \mid S)$. Hence the ux_i span the space on the left of (11). Now assume that $\Sigma\, a_i u x_i \in L(\mathfrak{a}\mathfrak{q} \mid S)$; then $\Sigma\, a_i x_i \in L(\mathfrak{q} \mid S)$, hence $v_{\mathfrak{q}}(\Sigma\, a_i x_i) > 0$ and since the $\bar{x}_i$ form a basis in $k_{\mathfrak{q}}$ we have $a_1 = \dots = a_f = 0$. This shows the ux_i to be linearly independent (mod $\mathfrak{a}\mathfrak{q} \mid S$). So we have established (11); now (10) follows by an easy induction.■

We can now establish the finiteness of $\dim_k L(\mathfrak{a})$:

Theorem 1.7

For any divisor $\mathfrak{a}$ on a function field K/k the space $L(\mathfrak{a})$ is finite-dimensional over k. If $\dim_k L(\mathfrak{a}) = l(\mathfrak{a})$, then

$$l(\mathfrak{a}) + d(\mathfrak{a}) \leqslant l(\mathfrak{b}) + d(\mathfrak{b}) \quad \textit{whenever } \mathfrak{a} \mid \mathfrak{b}. \tag{12}$$

Proof

Let S be the set of all prime divisors occurring in $\mathfrak{a}$ or $\mathfrak{b}$. Then S is finite and for $\mathfrak{a} \mid \mathfrak{b}$,

$$L(\mathfrak{b}) = L(\mathfrak{a}) \cap L(\mathfrak{b} \mid S).$$

Hence by the second isomorphism theorem (parallelogram rule, cf. A.1, 9.1) and Remark 3 above,

$$L(\mathfrak{a})/L(\mathfrak{b}) = L(\mathfrak{a})/(L(\mathfrak{a}) \cap L(\mathfrak{b} \mid S)) \cong (L(\mathfrak{a}) + L(\mathfrak{b} \mid S))/L(\mathfrak{b} \mid S)$$
$$\subseteq L(\mathfrak{a} \mid S)/L(\mathfrak{b} \mid S)$$

Here the right-hand side equals $d(\mathfrak{b}) - d(\mathfrak{a})$ by Lemma 1.6, hence

$$\dim_k(L(\mathfrak{a})/L(\mathfrak{b})) \leqslant d(\mathfrak{b}) - d(\mathfrak{a}). \tag{13}$$

Now for any divisor $\mathfrak{a}$ we can find a multiple $\mathfrak{b}$ which is integral and $\neq \mathfrak{e}$. As we have seen, for such $\mathfrak{b}$ we have $L(\mathfrak{b}) = 0$. Choosing $\mathfrak{b}$ in this way we see from (13) that $L(\mathfrak{a})$ is finite-dimensional. Thus, we obtain from (13),

$$l(\mathfrak{a}) - l(\mathfrak{b}) \leqslant d(\mathfrak{b}) - d(\mathfrak{a}),$$

and this yields (12) by rearrangement.■

Exercises

1. Show that the HCF of any finite set of divisors is integral if and only if each of the divisors is integral.
2. Show that for any divisor $\mathfrak{a}$, $L(\mathfrak{a}) = 0$ if $d(\mathfrak{a}) > 0$. Deduce that for any divisor $\mathfrak{a} \neq \mathfrak{e}$, $l(\mathfrak{a})$ is bounded by the degree of the denominator of $\mathfrak{a}$.
3. Show that for any integral divisor $\mathfrak{a}$ in a rational function field $k(x)$, $l(\mathfrak{a}^{-1}) + d(\mathfrak{a}^{-1}) = 1$.
4. Show that $L(\mathfrak{a} \mid S \cup T) = L(\mathfrak{a} \mid S) \cap L(\mathfrak{a} \mid T)$, for any sets S, T of places.
5. Show that for $\mathfrak{a} \mid \mathfrak{b}$ and for any S, finite or infinite,
$$\dim (L(\mathfrak{a} \mid S)/L(\mathfrak{b} \mid S)) \leqslant d(\mathfrak{b}) - d(\mathfrak{a}).$$
6. Let S be an infinite set of places on a rational function field $k(x)$. Give an example of a divisor $\mathfrak{a}$ such that $L(\mathfrak{a} \mid S)$ is infinite-dimensional.
7. Show that in any algebraic function field K/k, the field of constants has finite degree over k.
8. Let K/k be an algebraic function field (in one variable, with perfect ground field k). Show that the relative algebraic closure of k in K is the precise subfield of K on which every valuation of K over k is trivial.
9. Let K/k be an algebraic function field, $x, y \in K\backslash k$ and $f(x, y) = 0$ an irreducible equation for x, y over k. Show that for any $\alpha, \beta \in k$ satisfying $f(\alpha, \beta) = 0$ there is a place $\mathfrak{p}$ of K which is a common zero of $x - \alpha$ and $y - \beta$. (*Hint*: Take a suitable extension of the place of $k(x)$ corresponding to $x - \alpha$.)
10. Show that any automorphism of the rational function field $k(x)$ over k (i.e. leaving k elementwise fixed) has the form $x \mapsto (ax + b)/(cx + d)$, where $a, b, c, d, \in k$ are such that $ad - bc \neq 0$.

11. Let K/k be an algebraic function field, with a separating element x, and let $y \in K$ have the minimal equation over $k(x)$:
$$y^n + a_1 y^{n-1} + \ldots + a_n = 0.$$
Show that the divisor of y is the HCF of the functions $a_1, a_2^{1/2}, \ldots, a_n^{1/n}$.
12. Let K/k be a function field and $u, v \in K$. Show that if $u \neq v$, then there is a place on K, not a pole of u or v, at which u and v have different values.

4.2 PRINCIPAL DIVISORS AND THE DIVISOR CLASS GROUP

Let K/k be a function field. With each element u of $K^\times$ we can associate a divisor
$$(u) = \prod \mathfrak{p}^{v_\mathfrak{p}(u)}, \tag{1}$$
called a *principal divisor*. That we have a divisor follows because $v_\mathfrak{p}(u) = 0$ for almost all $\mathfrak{p}$; moreover, by the product formula, $d((u)) = 0$. The element u is determined by its divisor up to a constant factor, for if u, u' have the same divisor, then (u/u') is the unit divisor $\mathfrak{e}$, and so u/u' as an element without zeros must be a constant, by Proposition 1.5. Thus we can determine the function corresponding to a given divisor by prescribing its value at a given place not a zero or pole.

To examine (u) more closely, let us write it as
$$(u) = \frac{(u)_0}{(u)_\infty},$$
where $(u)_0$, $(u)_\infty$ are coprime integral divisors. The divisor $(u)_0$ is called the *numerator* or *divisor of zeros* of u, $(u)_\infty$ is the *denominator* or *divisor of poles* of u. It is clear that u^{-1} has numerator $(u)_\infty$ and denominator $(u)_0$. Explicitly we have
$$(u)_0 = \Pi\,\{\mathfrak{p}^{v_\mathfrak{p}(u)} \mid v_\mathfrak{p}(u) > 0\}, \qquad (u)_\infty = \Pi\,\{\mathfrak{p}^{-v_\mathfrak{p}(u)} \mid v_\mathfrak{p}(u) < 0\}.$$

Examples

(i) $k = \mathbf{C}, K = \mathbf{C}(t)$; write p_a for the place defined by $t - a$ and p_∞ for the place at ∞, defined by $1/t$. Then for $u = (t-a)/(t-b)^2$ we have $(u)_0 = p_a p_\infty, (u)_\infty = p_b^2$. Generally, a rational function in t, with numerator of degree r and denominator of degree s, has a zero of order $s - r$ at ∞ if $s > r$ and a pole of order $r - s$ for $r > s$, so the total degree is 0, as we know it must be.

(ii) $K = k(x, y)$, where $x^2 + y^2 = 1$. Consider the function $u = 2x/(x + y - 1)$; its numerator is $(0, -1)$ and denominator is (∞, ∞), where each place is represented by the values of x and y. We note that in the given representation the numerator and denominator both vanish at the place (0, 1). A different representation not involving this place would be $u = (x + y + 1)/y$; here both numerator and denominator vanish at $(-1, 0)$, and there is no representation of u as f/g, where f, g have coprime divisors of zeros.

Our first objective is to determine $d((u)_0)$; at the same time we shall give another proof that (u) is a divisor and has degree zero.

Theorem 2.1
Let K/k be a function field. Then for any $u \in K \backslash k$,

$$d((u)_0) = d((u)_\infty) = [K : k(u)]. \tag{2}$$

Proof
Let us write $(u)_0 = \mathfrak{n}$ for short. By Lemma 1.6 we have

$$\dim_k(L(\mathfrak{e} \mid S)/L(\mathfrak{n} \mid S)) = d(\mathfrak{n}),$$

where S is the set of all places occurring in $\mathfrak{n}$. The extension $K/k(u)$ is finite by definition; let $[K : k(u)] = N$ and choose any $N + 1$ elements $y_0, \ldots, y_N$ in $L(\mathfrak{e} \mid S)$. These elements are integral at S and by definition of N are linearly dependent over $k(u)$; thus there exist $f_i \in k(u)$ $(i = 0, \ldots, N)$, not all zero, such that

$$\sum f_i(u) y_i = 0.$$

Clearing denominators, we may assume that $f_i(u) \in k[u]$ and by omitting common factors we can ensure that the $f_i(0)$ are not all 0, say $f_i = \alpha_i + u g_i(u)$. Then

$$\sum \alpha_i y_i = -u \,.\, \sum g_i(u) y_i,$$

hence for any $\mathfrak{p} \in S$, $v_\mathfrak{p}(\Sigma \alpha_i y_i) \geqslant v_\mathfrak{p}(u)$, i.e. $\Sigma \alpha_i y_i \in L(\mathfrak{n} \mid S)$. Hence the y's are linearly dependent mod $L(\mathfrak{n} \mid S)$ over k and it follows that $d(\mathfrak{n}) \leqslant N$. By applying the argument to u^{-1} we find that $d((u)_\infty) \leqslant N$.

In the other direction, to show that N is bounded by $d(\mathfrak{n})$, let us take a basis $y_1, \ldots, y_N$ of K over $k(u)$. We may further assume the y_i to be integral over $k[u]$. Then for any $t \geqslant 0$, the elements

$$u^i y_j \quad (i = 0, \ldots, t,\ j = 1, \ldots, N)$$

are linearly independent over k. Now for any place $\mathfrak{p}$, $u \in \mathfrak{o}_\mathfrak{p} \Rightarrow y_j \in \mathfrak{o}_\mathfrak{p}$, hence any pole of any y_j is a pole of u and it follows, on writing $\mathfrak{d} = (u)_\infty$, that for each $j = 1, \ldots, N$ there exists $s_j \in \mathbf{N}$ such that $\mathfrak{d}^{s_j}(y_j)$ is integral. Put $s = \max\{s_1, \ldots, s_N\}$; then $\mathfrak{d}^s(y_j)$ is integral for $j = 1, \ldots, N$ and so of course is $\mathfrak{d}.(u)$. Hence for any t,

$$\mathfrak{d}^{s+t}\,.\,(u^i y_j) \text{ is integral,} \quad \text{for } i = 0, \ldots, t, j = 1, \ldots, N.$$

This means that $u^i y_j \in L(\mathfrak{d}^{-s-t})$ and the linear independence shows that

$$l(\mathfrak{d}^{-s-t}) \geqslant (t+1)N. \tag{3}$$

By Theorem 1.7, since $\mathfrak{d}^{-s-t} \mid \mathfrak{e}$, we have

$$l(\mathfrak{d}^{-s-t}) \leqslant l(\mathfrak{e}) + d(\mathfrak{e}) - d(\mathfrak{d}^{-s-t}) = 1 + 0 + (s+t)d(\mathfrak{d}).$$

Thus we have $l(\mathfrak{d}^{-s-t}) \leqslant (s+t)d(\mathfrak{d}) + 1$, and combining this with (3) we find

$$d(\mathfrak{d}) \geqslant \frac{t+1}{t+s}N - \frac{1}{t+s},$$

for fixed s and all t. Letting $t \to \infty$, we find that $d(\mathfrak{d}) \geqslant N$.
This shows that $d(\mathfrak{d}) = N$ and applied to u^{-1} the argument shows that $d(\mathfrak{n}) = N$.■

By (2) and (3) we have:

Corollary 2.2
For any u in a function field K, with divisor of poles $\mathfrak{d}$, there exists $s \in \mathbf{N}$ such that for all $t \in \mathbf{N}$,

$$l(\mathfrak{d}^{-s-t}) \geqslant (t+1)d(\mathfrak{d}).■ \tag{4}$$

We now return to the spaces $L(\mathfrak{a})$ and construct a canonical basis for $L(\mathfrak{a})$ which will incidentally provide another proof that these spaces are finite-dimensional.

Let K/k again be a function field and x any separating element in K. Write $A = k[x]$ and let B be the integral closure of A in K. Next, writing ∞ for the divisor of poles of x, let A' be the ring of all ∞-integral elements of $k(x)$, i.e. all $u \in k(x)$ such that $v_\infty(u) \geqslant 0$, and let B' be the integral closure of A' in K. Explicitly, A' consists of all rational functions of x whose numerator has degree at most equal to that of the denominator, i.e. functions that are finite at ∞, while B' consists of all $u \in K$ with $v_\mathfrak{p}(u) \geqslant 0$ for all $\mathfrak{p} \mid \infty$. We remark that $A \cap A' = B \cap B' = k$, by Proposition 1.5.

Lemma 2.3
Let K/k be a function field with separating element x. Given any divisor $\mathfrak{a}$, where $v_\mathfrak{p}(\mathfrak{a}) = \alpha_\mathfrak{p}$ for any place $\mathfrak{p}$, define

$$T = L(\mathfrak{a}^{-1})_x = \{u \in K \mid v_\mathfrak{p}(u) + \alpha_\mathfrak{p} \geqslant 0 \text{ for all } \mathfrak{p} \nmid \infty\},$$

$$U = L(\mathfrak{a}^{-1})_\infty = \{u \in K \mid v_\mathfrak{p}(u) + \alpha_\mathfrak{p} \geqslant 0 \text{ for all } \mathfrak{p} \mid \infty\},$$

where ∞ is the divisor of poles of x. If $n = [K : k(x)]$, then the space T has a basis $t_1, \ldots, t_n$ over A and U has a basis $u_1, \ldots, u_n$ over A' such that

$$u_i = t_i x^{e_i}, \quad e_i \in \mathbf{Z}, \ i = 1, \ldots, n. \tag{5}$$

Moreover, $T \cap U = L(\mathfrak{a}^{-1})$.

Proof

The last assertion is clear from the definitions. It remains to construct the bases. For any $a \in T, a \neq 0$, there exists a largest e such that $ax^e \in U$, and $e \geqslant 0$ if and only if $a \in T \cap U$. In fact these exponents e have an upper bound depending only on $\mathfrak{a}$, not on a. For if $a \in T$, and $ax^e \in U$, then

$$v_\mathfrak{p}(a) \geqslant -\alpha_\mathfrak{p} \quad \text{for } \mathfrak{p} \nmid \infty, \qquad v_\mathfrak{p}(a) + v_\mathfrak{p}(x^e) \geqslant -\alpha_\mathfrak{p} \quad \text{for } \mathfrak{p} \mid \infty. \tag{6}$$

Now we have

$$\sum_{\mathfrak{p} \mid \infty} f_\mathfrak{p} v_\mathfrak{p}(x^e) = n v_\infty(x^e) = -ne, \qquad \Sigma_\mathfrak{p} f_\mathfrak{p} v_\mathfrak{p}(a) = 0, \qquad \Sigma f_\mathfrak{p} \alpha_\mathfrak{p} = f_\mathfrak{a}.$$

Hence, multiplying (6) by $f_\mathfrak{p}$ and adding, we obtain

$$-ne \geqslant -f_\mathfrak{a}, \quad \text{so} \quad e \leqslant \frac{1}{n} f_\mathfrak{a}.$$

Let us call e the *exponent* of $\mathfrak{a}$ in T over U. We now choose $t_1, \ldots, t_n \in T$ as follows: $t_i \in T$ is chosen to be linearly independent of $t_1, \ldots, t_{i-1}$ over $k(x)$ and of maximal exponent e_i over U. From the definition we have

$$e_1 \geqslant \ldots \geqslant e_n. \tag{7}$$

Put $u_i = t_i x^{e_i}$ for $i = 1, \ldots, n$; then the u_i lie in U and like the t_i they are linearly independent over $k(x)$. We claim that the t_i form an A-basis for T and the u_i form an A'-basis for U.

Take $c \in T$ and write

$$c = \sum \gamma_i t_i, \quad \text{where } \gamma_i \in k(x).$$

We have to show that $\gamma_i \in A$. Write $\gamma_i = a_i + b_i/g$, where $a_i, b_i, g \in A$, g is a common denominator and $\deg b_i < \deg g$. We have

$$c' = c - \sum a_i t_i = \sum b_i t_i/g \in T.$$

If the b_i are not all 0, let the last non-zero one be b_r. Then

$$c' = \sum_1^r b_i t_i/g$$

is linearly independent of $t_1, \ldots, t_{r-1}$ and we have

$$c' x^{e_r + 1} = \sum_1^r x b_i x^{e_r} t_i/g = \sum_1^r x b_i u_i x^{e_r - e_i}/g.$$

Since $\deg b_i < \deg g$, we have $xb_i/g \in A'$; further, $e_r \leqslant e_i$ and so $u_i x^{e_r - e_i} \in U$, but this contradicts the choice of t_r. Hence $b_i = 0$ for all i and the t_i form indeed an A-basis of T.

To show that the u_i form an A'-basis of U, take $c \in U$ and write

$$c = \Sigma \gamma_i u_i, \quad \text{where } \gamma_i \in k(x).$$

We have to show that $\gamma_i \in A'$. Let us write $\gamma_i/x = a_i + b_i/xg$, where $a_i, b_i, g \in A'$, g is a common denominator of all the γ_i (but the division is by xg) and $\deg b_i \leqslant \deg g$. We have $\gamma_i = xa_i + b_i/g$, hence

$$c' = c - \sum b_i u_i/g = \sum x a_i u_i \in U,$$

and we must show that the a_i vanish. If this is not the case, let the last non-zero one be a_r. We have

$$x^{-e_r-1}c' = \sum_1^r a_i u_i x^{-e_r} = \sum_1^r a_i x^{e_i - e_r} t_i;$$

this is an element of T, linearly independent of $t_1, \ldots, t_{r-1}$ and with exponent over U at least $e_r + 1$. This again contradicts the choice of t_i; hence the u_i form a basis for U over A'.■

This result quickly leads to a formula for $l(\mathfrak{a})$:

Corollary 2.4
With the notations of Lemma 2.3 we have for any divisor $\mathfrak{a}$,

$$l(\mathfrak{a}^{-1}) = \sum (e_i + 1), \tag{8}$$

where the summation is extended over the non-negative e's in (7).

Proof
We have $L(\mathfrak{a}^{-1}) = T \cap U$. Consider any element of T:

$$c = \sum g_i t_i = \sum g_i x^{-e_i} u_i, \quad g_i \in A.$$

This element lies in U if and only if $g_i x^{-e_i} \in A'$, i.e. $\deg g_i \leqslant e_i$. So there are no such elements for $e_i < 0$, while for $e_i \geqslant 0$ the g_i satisfying this condition form an $(e_i + 1)$-dimensional k-space, hence (8) follows.■

Let us return to the divisor group D. The map $x \mapsto (x)$ is clearly a homomorphism from $K^\times$ to D. The kernel consists of the elements without poles or zeros, i.e. the constants (Proposition 1.5), while the image is the subgroup of principal divisors. This gives the second row of the diagram below; the first row arises by recalling that the principal divisors have degree 0. Here D_0 is the group of divisors of degree 0, C is the group of divisor classes and J the group of divisor classes of degree zero.

In contrast to the situation in number theory the divisor class group C here is infinite. This is because we have used all the divisors; to obtain a Dedekind

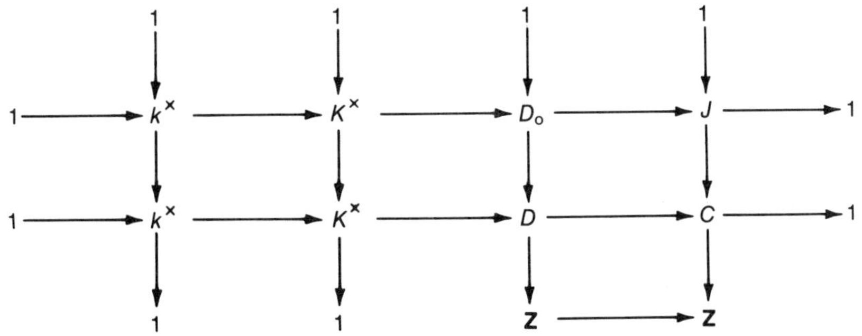

ring and a finite divisor class group we must leave out at least one prime divisor (so that Theorem 3.4.1 can be applied). On the other hand, in the present situation even J is infinite, for an infinite ground field. When $k = \mathbf{C}$, it is a compact abelian Lie group, called the *Jacobian variety*. In section 4.8 we shall show that $J \cong \mathbf{C}^g/\Omega$, where g is the genus of the function field and Ω is the period lattice (Abel–Jacobi theorem).

Let C be a divisor class. Given $\mathfrak{a} \in C$, any other divisor $\mathfrak{a}' \in C$ has the form $\mathfrak{a}' = \mathfrak{a} \,.\, (x)$ for some $x \in K^\times$. Hence

$$d(\mathfrak{a}') = d(\mathfrak{a}) + d(x) = d(\mathfrak{a}).$$

We can therefore define the *degree* of the divisor class C unambiguously as $d(C) = d(\mathfrak{a})$. Similarly we have $l(\mathfrak{a}) = l(\mathfrak{a}')$, for the map $y \mapsto xy$ provides an isomorphism $L(\mathfrak{a}) \cong L(\mathfrak{a}')$.

As we have noted earlier, $l(\mathfrak{a}) = 0$ for any integral divisor $\mathfrak{a} \neq \mathfrak{e}$, so $l(\mathfrak{a}^{-1})$ provides in some sense a better measure; we shall define, for any divisor class C,

$$N(C) = l(\mathfrak{a}^{-1}), \quad \text{where } \mathfrak{a} \in C. \tag{9}$$

Thus $N(C)$ gives the dimension of the space of all functions with poles confined to $\mathfrak{a}$.

Since any integral divisor $\mathfrak{a}$ satisfies $\mathfrak{a}^{-1} \mid \mathfrak{e}$ and $l(\mathfrak{e}) + d(\mathfrak{e}) = 1 + 0 = 1$, we have, by Theorem 1.7, $l(\mathfrak{a}^{-1}) + d(\mathfrak{a}^{-1}) \leqslant 1$. This proves:

Proposition 2.5
Let K/k be any function field and C a divisor class. If C contains an integral divisor, then

$$N(C) \leqslant d(C) + 1. \blacksquare \tag{10}$$

We conclude the section by noting a geometrical interpretation of divisor classes. Let $\mathfrak{a}$ be a divisor in an algebraic function field K/k and write $[\mathfrak{a}]$ for the corresponding divisor class. This class may be regarded as a projective k-space: any point of $[\mathfrak{a}]$ is a divisor equivalent to $\mathfrak{a}$, hence of the form $u\mathfrak{a}$, where $u \in K^\times$ is determined up to a factor in k. We can form sums of points:

thus

$$\lambda_1(u_1\mathfrak{a}) + \lambda_2(u_2\mathfrak{a}) = (\lambda_1 u_1 + \lambda_2 u_2)\mathfrak{a} \quad \text{where } u_i \in K^{\times},\ \lambda_i \in k^{\times}.$$

For a fixed divisor $\mathfrak{a}$, any space of divisors $u\mathfrak{a}$ with $u \equiv 0 \pmod{\mathfrak{a}^{-1}}$ is called a *linear series* and the space of all such divisors is a *complete linear series*; by Theorem 1.7 it has finite dimension $l(\mathfrak{a}^{-1}) - 1$. When $\mathfrak{a}$ is an integral divisor, the corresponding linear series is said to be *effective*; it corresponds to a finite set of points (with multiplicities) on the defining curve $f = 0$ of K. For example, on a straight line, the set of all n-tuples of points is a complete linear series; the same holds for a circle, because $k(x, y)$, where $x^2 + y^2 = 1$, is isomorphic to $k(t)$, where $t = y/x$, $x = (1 - t^2)/(1 + t^2)$, $y = 2t/(1 + t^2)$. This is expressed by saying that a circle is birationally equivalent to a straight line. On the other hand, as we shall see later, the field defined by $x^3 + y^2 = 1$ is not rational and its complete series have a different form, which we shall meet in section 4.7.

Exercises

1. Show that the set of functions with poles in a given integral divisor $\mathfrak{a}$ depends linearly on $d(\mathfrak{a})$ parameters. Hence prove Proposition 2.5 by showing that if $l(\mathfrak{a}^{-1}) > d(\mathfrak{a}) + 1$, then there are two linear combinations without poles and hence a non-constant function without poles.
2. Consider the field $k(x, y)$, where k is of characteristic $\neq 2, 3$ and x, y satisfy $x(x - 1)(x - 2) + y^3 = 1$. Find the divisors of poles and zeros of x and y.
3. An integral divisor $\mathfrak{a}$ is said to be *isolated* if there is no other integral divisor equivalent to $\mathfrak{a}$ (i.e. in the same divisor class). Show that this is so precisely when $l(\mathfrak{a}^{-1}) = 1$.
4. The HCF of all integral divisors in a divisor class C is called the *divisor* of C, and C is said to be *primitive* if this divisor is $\mathfrak{e}$. Show that a divisor $\mathfrak{a}^{-1}$ can be written as an HCF of finitely many elements of K if and only if the divisor class C containing $\mathfrak{a}$ has positive dimension and is primitive. Show that, moreover, $d(C) \geqslant 0$.
5. Show that the divisor of a divisor class C is always isolated (cf. Exercises 3 and 4) and may be characterized as the greatest integral divisor $\mathfrak{d}$ such that $N(C\mathfrak{d}^{-1}) = N(C)$. Show further that C is primitive if and only if $N(C\mathfrak{p}^{-1}) < N(C)$ for all prime divisors $\mathfrak{p}$.
6. Let $K = k(x, y)$ be a function field over k. Given $a, b \in k$, write $(x - a) = \mathfrak{p}_1^{e_1} \ldots \mathfrak{p}_s^{e_s}$ and let $\mathfrak{p}_1, \ldots, \mathfrak{p}_t$ $(0 \leqslant t \leqslant s)$ be the places at which $x = a$, $y = b$. If $f(x, y) = 0$ is the irreducible equation defining K, show that $f(a, y)$ has a zero at $y = b$ of order $e_1 + \ldots + e_t$.

   Deduce that y is an integral function of x if and only if the denominator of (y) contains only prime divisors occurring in the denominator of (x).

4.3 RIEMANN'S THEOREM AND THE SPECIALTY INDEX

We saw from the remark before Proposition 2.5 that for an integral divisor $\mathfrak{a}$, $l(\mathfrak{a}^{-1})$ is bounded above: $l(\mathfrak{a}^{-1}) \leqslant d(\mathfrak{a}) + 1$. For non-integral $\mathfrak{a}$ this need not hold because then $d(\mathfrak{a})$ may be negative. We shall now find a lower bound for $l(\mathfrak{a}^{-1})$; more precisely, we give a lower bound for $l(\mathfrak{a}) + d(\mathfrak{a})$ which will hold for all $\mathfrak{a}$, integral or not.

Theorem 3.1 (*Riemann's theorem*)
Let K/k be a function field. Then as $\mathfrak{a}$ runs over all divisors, $l(\mathfrak{a}) + d(\mathfrak{a})$ is bounded below. Thus there exists an integer g such that

$$l(\mathfrak{a}) + d(\mathfrak{a}) \geqslant 1 - g. \tag{1}$$

Moreover, given any $x \in K \backslash k$, the lower bound is attained for $\mathfrak{a} = (x)_\infty^{-m}$, for all large m.

The constant g in (1) is an important invariant of K, called the *genus*. It can assume any non-negative integer value, and it has geometrical and function-theoretic interpretations, some of which we shall meet later.

Proof
Given $x \in K \backslash k$, let $\mathfrak{d} = (x)_\infty$ be its denominator. By Corollary 2.2 there exists $s \in \mathbf{N}$ such that for all $t \geqslant 0$,

$$l(\mathfrak{d}^{-s-t}) \geqslant (t+1)d(\mathfrak{d}).$$

Write $m = s + t$, so that for $m \geqslant s$,

$$l(\mathfrak{d}^{-m}) + d(\mathfrak{d}^{-m}) \geqslant (t + 1 - m)d(\mathfrak{d}).$$

Hence we have

$$l(\mathfrak{d}^{-m}) + d(\mathfrak{d}^{-m}) \geqslant (1-s)d(\mathfrak{d}), \tag{2}$$

and here the right-hand side is independent of m. If $m < s$, then $\mathfrak{d}^{-s} \mid \mathfrak{d}^{-m}$ and so, by Theorem 1.7, the same inequality holds in this case. Thus (2) holds for all $m \in \mathbf{Z}$, and this shows that the left-hand side of (2) has a lower bound as m varies. Let us denote this lower bound by $1 - g$, where g may still depend on $\mathfrak{d}$. By Proposition 2.5, g is a non-negative integer and by Theorem 1.7, if equality holds for some m, it also holds for all larger values.

Now take any integral divisor $\mathfrak{a}$. Then for any m, with $\mathfrak{d} = (x)_\infty$ as before, we have

$$l(\mathfrak{d}^{-m}\mathfrak{a}) + d(\mathfrak{d}^{-m}\mathfrak{a}) \geqslant l(\mathfrak{d}^{-m}) + d(\mathfrak{d}^{-m}) \geqslant 1 - g.$$

It follows that

$$\begin{aligned} l(\mathfrak{d}^{-m}\mathfrak{a}) &\geqslant -d(\mathfrak{d}^{-m}\mathfrak{a}) + 1 - g \\ &= md(\mathfrak{d}) - d(\mathfrak{a}) + 1 - g. \end{aligned}$$

Further, $d(\mathfrak{d}) > 0$ because x has at least one pole, so the right-hand side is positive for large enough m. For such an m we can find $z \in L(\mathfrak{d}^{-m}\mathfrak{a}), z \neq 0$. Then $(z)\,\mathfrak{d}^m\mathfrak{a}^{-1}$ is integral, i.e. $\mathfrak{d}^{-m} \mid (z)\mathfrak{a}^{-1}$ and so, recalling that l and d are class functions, we find that

$$\begin{aligned} l(\mathfrak{a}^{-1}) + d(\mathfrak{a}^{-1}) &= l((z)\mathfrak{a}^{-1}) + d((z)\mathfrak{a}^{-1}) \\ &\geqslant l(\mathfrak{d}^{-m}) + d(\mathfrak{d}^{-m}) \geqslant 1 - g. \end{aligned}$$

This holds for any integral divisor $\mathfrak{a}$. To establish the result generally, let $\mathfrak{a} = \mathfrak{a}_1\mathfrak{a}_2^{-1}$, where $\mathfrak{a}_1, \mathfrak{a}_2$ are integral. Then $\mathfrak{a}_2^{-1} \mid \mathfrak{a}$ and so

$$l(\mathfrak{a}) + d(\mathfrak{a}) \geqslant l(\mathfrak{a}_2^{-1}) + d(\mathfrak{a}_2^{-1}) \geqslant 1 - g.$$

This shows that $1 - g$ is the required lower bound, and so is independent of the choice of x.■

As a consequence we have:

Corollary 3.2
In a function field of genus g let $\mathfrak{a}$ be any integral divisor of degree $g+1$. Then there is a non-constant function with poles confined to $\mathfrak{a}$.

For by Theorem 3.1, $l(\mathfrak{a}^{-1}) \geqslant d(\mathfrak{a}) + 1 - g = 2$, hence $L(\mathfrak{a}^{-1})$ contains a non-constant function.■

Let $\mathfrak{a}$ be any divisor in a function field K/k. The integer

$$\delta(\mathfrak{a}^{-1}) = l(\mathfrak{a}) + d(\mathfrak{a}) + g - 1, \tag{3}$$

which by Theorem 3.1 is non-negative, is called the *index of specialty* of $\mathfrak{a}$. We say that $\mathfrak{a}$ is *special* if $\delta(\mathfrak{a}^{-1}) > 0$, *non-special* if $\delta(\mathfrak{a}^{-1}) = 0$. We have also seen that for the divisor of poles $\mathfrak{d}$, of an element x, $\mathfrak{d}^{-m}$ is non-special for all large m (the case of a non-special divisor should be thought of as the norm, with some exceptional cases that are 'special'). Further, by Theorem 1.7, if $\mathfrak{a} \mid \mathfrak{b}$, then $\delta(\mathfrak{a}^{-1}) \leqslant \delta(\mathfrak{b}^{-1})$, hence if $\mathfrak{b}$ is non-special, then so is any divisor of $\mathfrak{b}$.

Our next problem is to identify $\delta(\mathfrak{a}^{-1})$. Since this is a non-negative integer, it seems natural to look for a space whose dimension is $\delta(\mathfrak{a}^{-1})$. For this purpose it is convenient to introduce another construction. Let Σ be the set of all places on K and consider the direct power K^Σ, as a ring. By an *adele*† one understands an element $x = (x_\mathfrak{p})$ of K^Σ such that $v_\mathfrak{p}(x_\mathfrak{p}) \geqslant 0$ for almost all $\mathfrak{p} \in \Sigma$. It is clear that the set Λ of all adeles is a subring of K^Σ. Moreover, Λ is a K-space in which K is embedded by the diagonal map $x \mapsto (x_\mathfrak{p})$, where $x_\mathfrak{p} = x$ for all $\mathfrak{p}$; this is so because each element of K has only finitely many poles. The elements in the image of K in Λ are called the *principal adeles*. We remark that the ring Λ is vast compared with K, but it has the advantage of

† Usually an adele is defined as an element of $\prod K_\mathfrak{p}$, where $K_\mathfrak{p}$ is the completion of K at $\mathfrak{p}$, but the above form is sufficient for our purpose.

being easier to work with, as is shown e.g. by Proposition 3.3 below. Each valuation $v_{\mathfrak{p}}$ may be extended to Λ by writing $v_{\mathfrak{p}}(x) = v_{\mathfrak{p}}(x_{\mathfrak{p}})$, and we shall write $x \equiv 0 \pmod{\mathfrak{a}}$ to mean: $v_{\mathfrak{p}}(x) \geqslant v_{\mathfrak{p}}(\mathfrak{a})$ for all $\mathfrak{p}$.

For each divisor $\mathfrak{a}$ we define a subspace of Λ by the rule

$$\Lambda(\mathfrak{a}) = \{x \in \Lambda \mid x \equiv 0 \pmod{\mathfrak{a}}\}.$$

Each $\Lambda(\mathfrak{a})$ is a k-space (although Λ itself is a K-space). It is clear that

$$L(\mathfrak{a}) = \Lambda(\mathfrak{a}) \cap K,$$

and this leads to a relation between the dimensions of the $\Lambda(\mathfrak{a})$ analogous to that of Lemma 1.6.

Proposition 3.3
In a function field K/k, let $\mathfrak{a}$, $\mathfrak{b}$ be divisors such that $\mathfrak{a} \mid \mathfrak{b}$. Then $\Lambda(\mathfrak{a}) \supseteq \Lambda(\mathfrak{b})$ and

$$\dim_k(\Lambda(\mathfrak{a})/\Lambda(\mathfrak{b})) = d(\mathfrak{b}) - d(\mathfrak{a}).$$

Proof
Let S be the set of all places occurring in $\mathfrak{a}$ or $\mathfrak{b}$ and define a map $\theta : L(\mathfrak{a} \mid S) \to \Lambda(\mathfrak{a})$ by the rule

$$(x\theta)_{\mathfrak{p}} = \begin{cases} x & \text{if } \mathfrak{p} \in S, \\ 0 & \text{if } \mathfrak{p} \notin S, \end{cases}$$

Clearly this is a k-linear map and since $\Lambda(\mathfrak{a}) \supseteq \Lambda(\mathfrak{b})$, we obtain a homomorphism

$$\phi : L(\mathfrak{a} \mid S) \to \Lambda(\mathfrak{a})/\Lambda(\mathfrak{b}), \tag{4}$$

by combining θ with the natural homomorphism to the quotient. The map (4) is surjective, for if $x \in \Lambda(\mathfrak{a})$, then there exists $x' \in K$ such that $v_{\mathfrak{p}}(x - x') \geqslant v_{\mathfrak{p}}(\mathfrak{b})$ for $\mathfrak{p} \in S$, so ϕ maps x' to the image of x. Further, the kernel of ϕ is $\Lambda(\mathfrak{b}) \cap L(\mathfrak{a} \mid S) = L(\mathfrak{b} \mid S)$, so we find that

$$\Lambda(\mathfrak{a})/\Lambda(\mathfrak{b}) \cong L(\mathfrak{a} \mid S)/L(\mathfrak{b} \mid S),$$

and the right-hand side has dimension $d(\mathfrak{b}) - d(\mathfrak{a})$ over k, by Lemma 1.6.■

We remark that the result remains unaffected if we had defined Λ in terms of $\prod K_{\mathfrak{p}}$. We can now identify a space whose dimension is $\delta(\mathfrak{a}^{-1})$.

Theorem 3.4
For any divisor $\mathfrak{a}$ of a function field K/k,

$$\dim_k(\Lambda/\Lambda(\mathfrak{a}) + K) = \delta(\mathfrak{a}^{-1}) = l(\mathfrak{a}) + d(\mathfrak{a}) + g - 1. \tag{5}$$

Moreover, if $\mathfrak{a} \mid \mathfrak{b}$, *then*

$$\dim_k [(\Lambda(\mathfrak{a}) + K)/(\Lambda(\mathfrak{b}) + K)] = l(\mathfrak{b}) + d(\mathfrak{b}) - l(\mathfrak{a}) - d(\mathfrak{a}). \tag{6}$$

Proof

It is clear that (6) is an immediate consequence of equation (5), but we shall prove (6) first. Since $\Lambda(\mathfrak{a}) \supseteq \Lambda(\mathfrak{b})$ and $\Lambda(\mathfrak{a}) \cap K = L(\mathfrak{a})$, we have, by the modular law,

$$(\Lambda(\mathfrak{b}) + K) \cap \Lambda(\mathfrak{a}) = \Lambda(\mathfrak{b}) + (K \cap \Lambda(\mathfrak{a})) = \Lambda(\mathfrak{b}) + L(\mathfrak{a}).$$

Hence

$$\frac{\Lambda(\mathfrak{a}) + K}{\Lambda(\mathfrak{b}) + K} \cong \frac{\Lambda(\mathfrak{a})}{(\Lambda(\mathfrak{b}) + K) \cap \Lambda(\mathfrak{a})} \cong \frac{\Lambda(\mathfrak{a})}{\Lambda(\mathfrak{b}) + L(\mathfrak{a})} \cong \frac{\Lambda(\mathfrak{a})/\Lambda(\mathfrak{b})}{(\Lambda(\mathfrak{b}) + L(\mathfrak{a}))/\Lambda(\mathfrak{b})}$$

$$\cong \frac{\Lambda(\mathfrak{a})/\Lambda(\mathfrak{b})}{L(\mathfrak{a})/L(\mathfrak{a}) \cap \Lambda(\mathfrak{b})}.$$

Since $L(\mathfrak{a}) \cap \Lambda(\mathfrak{b}) = L(\mathfrak{b})$, the last expression becomes

$$\frac{\Lambda(\mathfrak{a})/\Lambda(\mathfrak{b})}{L(\mathfrak{a})/L(\mathfrak{b})}.$$

Now (6) follows by Proposition 3.3 and the definition of l.

The relation (6) can be rewritten as

$$\dim_k ((\Lambda(\mathfrak{a}) + K)/(\Lambda(\mathfrak{b}) + K)) = \delta(\mathfrak{b}^{-1}) - \delta(\mathfrak{a}^{-1}) \leqslant \delta(\mathfrak{b}^{-1}). \tag{7}$$

Let us fix $\mathfrak{b}$, choose any non-special $\mathfrak{c}$ and take $\mathfrak{a}$ to satisfy $\mathfrak{a} \mid \mathfrak{b}$, $\mathfrak{a} \mid \mathfrak{c}$. Then $\mathfrak{a}$ is non-special, i.e. $\delta(\mathfrak{a}^{-1}) = 0$, and so by (7),

$$\dim_k (\Lambda/(\Lambda(\mathfrak{b}) + K)) \geqslant \delta(\mathfrak{b}^{-1}). \tag{8}$$

If this inequality were strict, we could find elements $x_1, \ldots, x_n \in \Lambda$ which are linearly independent (mod $\Lambda(\mathfrak{b}) + K$), where $n > \delta(\mathfrak{b}^{-1})$. We now choose $\mathfrak{a}$ to be non-special so that $\mathfrak{a} \mid \mathfrak{b}$ and $x_i \equiv 0 \pmod{\mathfrak{a}}$. Then

$$\dim_k ((\Lambda(\mathfrak{a}) + K)/(\Lambda(\mathfrak{b}) + K)) \geqslant n > \delta(\mathfrak{b}^{-1}),$$

which contradicts (7). So equality holds in (8) and (5) follows.■

An r-dimensional linear series associated with a divisor of degree n is usually denoted by g_n^r. Theorem 3.1 and the remark preceding it show that for every integral divisor $\mathfrak{a}$,

$$d(\mathfrak{a}) - g \leqslant l(\mathfrak{a}^{-1}) - 1 \leqslant d(\mathfrak{a}).$$

Hence for any complete linear series g_n^r we have $r \leqslant n$ with equality for some series whenever the function field is rational. This necessary condition for rationality can also be shown to be sufficient.

Exercises

1. Verify that a rational function field has genus zero and that every integral divisor is non-special.
2. Show that for an algebraic function field of genus at most 1, if d is the minimum of the degrees of the prime divisors, then every divisor has a degree divisible by d. (*Hint*: Let $\mathfrak{p}$ be a prime divisor of degree d and for any divisor $\mathfrak{a}$ apply Riemann's theorem to $\mathfrak{a}\mathfrak{p}^{-r}$ for suitable r, to find an element of $L(\mathfrak{a}\mathfrak{p}^{-r})$ and determine its divisor.)

4.4 THE GENUS

In section 4.3 we met the genus g of a function field K, defined as the least upper bound of the expression $1 + d(\mathfrak{a}) - l(\mathfrak{a}^{-1})$, for any divisor $\mathfrak{a}$. In this section we shall describe the Riemann surface corresponding to a given function field (in the case where the ground field is $\mathbf{C}$) and obtain a formula for the 'geometric genus', i.e. the number of handles of the surface, when expressed in normal form; later we shall see that this agrees with the genus as previously defined. There is also a purely geometric interpretation of the genus in terms of the singular points of the corresponding algebraic curve (the 'deficiency').

To begin with we shall consider the values taken by g in simple cases, beginning with rational fields.

Theorem 4.1

Any rational function field (of one variable) is of genus 0. Conversely, a function field K of genus 0 which has a place of degree 1 is rational.

Proof

Let $K = k(x)$ and denote by ∞ the divisor of poles of x. Then ∞ is of degree 1, and so by Theorem 3.1 we have

$$g = 1 - \min \{ l(\infty^{-m}) - m \},$$

and it remains to compute $l(\infty^{-m})$. If $u \in L(\infty^{-m})$, then the only pole of u is at ∞. Write $u = f/g$, where f, g are coprime polynomials in x. If g is not constant, it has a finite zero and so u has a pole other than ∞. This was not allowed, so $u = f$ is a polynomial, and the condition $u \equiv 0 \pmod{\infty^{-m}}$ requires the pole of u to have order at most m, i.e. $\deg f \leqslant m$. Thus the elements of $L(\infty^{-m})$ are the linear combinations of $1, x, \ldots, x^m$, so

$$l(\infty^{-m}) = m + 1, \qquad l(\infty^{-m}) - d(\infty^m) = 1$$

and hence $g = 0$.

Conversely, let K be a field of genus 0 and assume that there is a place $\mathfrak{p}$ of degree 1. By Theorem 3.1, $l(\mathfrak{p}^{-1}) - d(\mathfrak{p}) \geqslant 1$, hence $l(\mathfrak{p}^{-1}) \geqslant 2$. Thus there exists $x \in K$ such that $x \equiv 0 \pmod{\mathfrak{p}^{-1}}$ and $1, x$ are linearly independent over the ground field k. Hence the divisor of poles of x is $\mathfrak{p}$, which is of degree 1, therefore by Theorem 2.1, $[K : k(x)] = 1$ and this shows that $K = k(x)$.■

When the field of constants is algebraically closed, every place has degree 1 and Theorem 4.1 applies. In other cases there may be no place of degree 1, e.g. $\mathbf{R}(x, y)$, where $x^2 + y^2 + 1 = 0$ (cf. Exercise 1). However, it can be shown that every field of genus 0 has a place of degree 1 or 2 (cf. Exercise 4).

Consider next the case $g = 1$. A function field of genus 1 is also called an *elliptic* function field; this name will be explained in section 4.7, when we come to discuss such fields in more detail. We shall further assume that our field has a place of degree 1 (of course this is automatic when the ground field is algebraically closed).

Let K/k be an elliptic function field, with a place $\mathfrak{p}$ of degree 1. Then by Theorem 3.1, $l(\mathfrak{p}^{-2}) \geqslant d(\mathfrak{p}^2) = 2$; hence there exists $x \in K \backslash k$ such that $x \equiv 0 \pmod{\mathfrak{p}^{-2}}$. The divisor of poles of x is either $\mathfrak{p}$ or $\mathfrak{p}^2$; if it were $\mathfrak{p}$, then the argument of Theorem 4.1 would show that $K = k(x)$, and this is of genus 0. Hence the divisor of poles of x must be $\mathfrak{p}^2$, and by Theorem 2.1, K is a quadratic extension of $k(x)$.

Since $l(\mathfrak{p}^{-3}) \geqslant d(\mathfrak{p}^3) = 3$, the space $L(\mathfrak{p}^{-3})$ is at least three-dimensional and so contains an element y linearly independent of $1, x$. We claim that the divisor of poles of y is exactly $\mathfrak{p}^3$. For if not, then y would lie in $L(\mathfrak{p}^{-2})$, which has $1, x$ as basis, and so y would be linearly dependent on $1, x$, a contradiction. Hence y has $\mathfrak{p}^3$ as divisor of poles and $[K : k(y)] = 3$. Since the degrees of K over $k(x)$ and $k(y)$ are coprime, it follows that neither of $k(x), k(y)$ contains the other, and since both degrees are prime, we see that $K = k(x, y)$. Moreover, y satisfies a quadratic equation over $k(x)$. In section 4.6 we shall see that by a more precise analysis we can choose x, y so that y^2 is a polynomial of degree 3 in x over k. Conversely, if k is a field of characteristic not 2 and $K = k(x, y)$, where x is transcendental over k and $y^2 = \phi(x)$, ϕ being a polynomial of degree 3 with no multiple factor, then K is of genus 1. This is a consequence of the following more general result, which provides an upper bound for g.

Theorem 4.2

Let $K = k(x, y)$ be a function field over an infinite field k, in which the polynomial relation between x and y has total degree n. Then the genus g of K satisfies the relation

$$g \leqslant \frac{1}{2}(n-1)(n-2). \tag{1}$$

Proof

Let the equation relating x and y be

$$F(x, y) = 0, \tag{2}$$

and write F_n for the homogeneous part of degree n. If we put $x = x' + \alpha y$ and express F in terms of x' and y, then the coefficient of y^n in F_n is $F_n(\alpha, 1)$ and this is $\neq 0$ for some $\alpha \in k$. We may therefore change the variables so that F has a term in y^n. Let ∞ be the pole of x; then we see by looking at the dominant terms in (2) that

$$v_{\mathfrak{p}}(y) \begin{cases} \geqslant 0 & \text{for } \mathfrak{p} \nmid \infty, \\ \geqslant v_{\mathfrak{p}}(x) & \text{for } \mathfrak{p} \mid \infty. \end{cases}$$

Write $\mathfrak{a} = \infty^s$ and take s so large that $\mathfrak{a}$ is non-special. Then

$$l(\mathfrak{a}^{-1}) + d(\mathfrak{a}^{-1}) + g - 1 = 0,$$

and so, since $d(\infty) = n$, we find that

$$g = 1 + ns - l(\mathfrak{a}^{-1}). \tag{3}$$

It only remains to obtain an estimate for $l(\mathfrak{a}^{-1})$. The space $L(\mathfrak{a}^{-1})$ contains all linear combinations of $x^i y^j$ such that $i + j \leqslant s$. For fixed j this gives $s - j + 1$ values of i, so in all we have

$$\begin{aligned}(s+1) + s + (s-1) + \ldots + (s-n+2) &= n(s-n+2) + \frac{1}{2}n(n+1) \\ &= ns + 1 - \frac{1}{2}(n-1)(n-2).\end{aligned}$$

Inserting this lower bound for $l(\mathfrak{a}^{-1})$ in (3), we obtain (1).■

This result shows for example that every conic over an algebraically closed field is rational; however, as we shall see, non-rational cubics exist. We remark without proof that for an algebraic curve given by (2), with multiple points of multiplicities r_i the genus is given by

$$g = \frac{1}{2}(n-1)(n-2) - \frac{1}{2}\sum r_i(r_i - 1)$$

(cf. Semple and Roth (1949), p. 54).

Let us digress briefly to describe the geometrical significance of the genus. We shall assume that the reader has some acquaintance with topology, but the rest of this section will not be needed later and so may be skipped (or better, skimmed).

A *Riemann surface* is defined as a one-dimensional complex analytic manifold, usually taken to be connected. Here a *manifold* is understood to be a Hausdorff space with a family of local coordinate systems covering it, which in the case of a complex analytic manifold are related in their common domain by biholomorphic transformations (i.e. holomorphic transformations with an inverse, again holomorphic). As a topological space a Riemann surface is two-dimensional, because **C** is two-dimensional over **R**. Moreover, it is orientable: we can orient any neighbourhood by the sign of i; since coordinate transformations are biholomorphic, this gives a coherent orientation to the whole collection of neighbourhoods.

Let K be a function field with $\mathbf{C}$ as constant field and take $x \in K\backslash\mathbf{C}$. If $[K : \mathbf{C}(x)] = n$, K can be constructed as n-fold covering of the Riemann sphere $\mathbf{C}(x)$. The latter is a 2-sphere, coordinatized by projection from the north pole N. If the line NP cuts the tangent plane at the south pole in the point x, we take x as the complex coordinate of P and this is valid for every point other than N. Similarly, the coordinate $x' = x^{-1}$ gives a representation for all points other than the south pole.

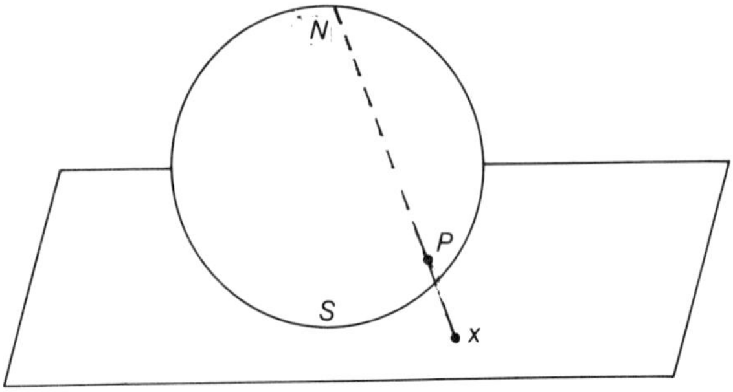

We have $K = \mathbf{C}(x, y)$, where y satisfies an equation of degree n over $\mathbf{C}(x)$. A point $\alpha \in \mathbf{C}$ gives in general n values for y; the exceptions are the branch points. Let the factorization of the principal divisor $(x - \alpha)$ be

$$(x - \alpha) = \prod \mathfrak{p}_i^{e_i};$$

then the branch points are the $\mathfrak{p}_i$ at which $e_i > 1$. If t_i is a uniformizer at $\mathfrak{p}_i$, we have

$$x = \alpha + \sum_{\nu = e_i}^{\infty} c_\nu t_i^\nu \quad (c_\nu \in \mathbf{C}).$$

When $e_i = 1$, we can take $x - \alpha$ as uniformizer, or at ∞, x^{-1}. The branch points are points at which the n sheets of the surface come together. We give some examples to describe the situation.

1. $y - x^2 = 0$. The branch points are 0, ∞. We have a two-sheeted covering, obtained by taking two spheres, cutting them along an arc from N to S and gluing them together along this arc, so as to cross over from one to the other along this cut. Topologically this is again a sphere (the y-sphere).
2. $(y^2 + 1)^3 = x$. This is a six-sheeted covering of the y-sphere, with branch points 0, 1, ∞. The discriminant has degree $\sum (e_i - 1) = 10$.

The permutations of the sheets induced at the branch points are $\gamma_0 = (123)(456)$, $\gamma_1 = (14)$, $\gamma_\infty = (123\,456)^{-1}$ and we have $\gamma_0\gamma_1\gamma_\infty = 1$.

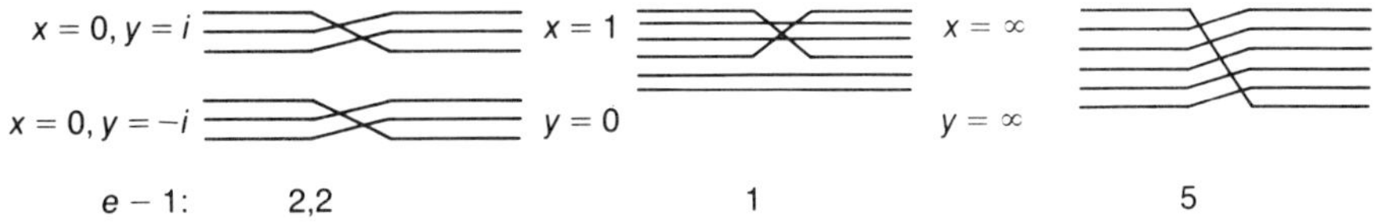

Generally, in an n-sheeted covering of the Riemann sphere, if the permutations induced on the sheets at the branch points are $\gamma_1, \ldots, \gamma_r$ then taken in a suitable order, we have $\gamma_1 \ldots \gamma_r = 1$. This leads to an interesting way of realizing groups as Galois groups. Any finite group G can be realized as a transitive permutation group of finite degree n, say. If G is generated by $\gamma_1, \ldots, \gamma_r$ satisfying the relation $\gamma_1 \ldots \gamma_r = 1$, then we can take n copies of the Riemann sphere, mark r points $P_1, \ldots, P_r$ in the same positions on each, with cuts between them and glue the n sheets together at each cut in accordance with the permutation γ_i at P_i. In this way one obtains an algebraic function field with G as a Galois group; this is in contrast to the situation in number theory, where it is still not known whether for every finite group G there is a Galois extension of $\mathbf{Q}$ with group G (cf. Matzat (1987)).

We recall from topology the classification of closed orientable surfaces (cf. e.g. Lefschetz (1949)). The Riemann surface of a function field, besides being orientable, is also compact (as finite covering surface of the sphere). Now every compact orientable two-dimensional manifold is homeomorphic to a sphere with p handles, where $p \geqslant 0$. By cutting it along $2p$ arcs $a_1, \ldots, a_p$, $b_1, \ldots, b_p$ we may describe it as a $2p$-gon with the boundary

$$a_1 b_1 a_1^{-1} b_1^{-1} a_2 b_2 a_2^{-1} b_2^{-1} \ldots a_p b_p a_p^{-1} b_p^{-1}.$$

Examples

$p = 0$ is the sphere

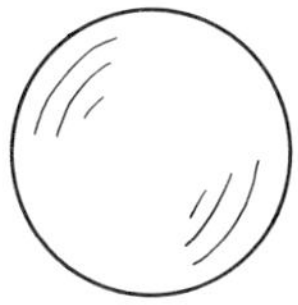

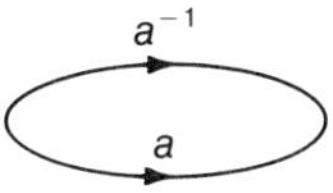

$p = 1$ is the torus

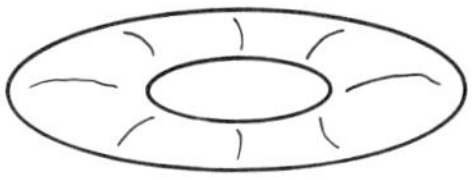

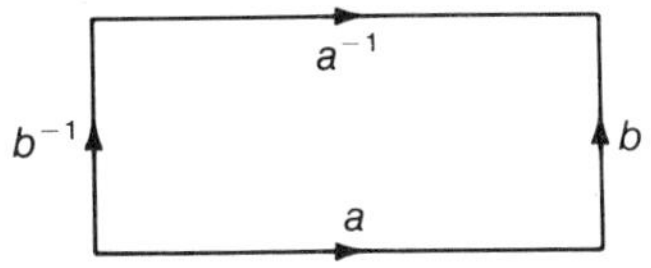

Generally we have a surface with p holes or handles. This number can be obtained as follows. Every Riemann surface can be triangulated and for any triangulation of the surface we have the Euler formula:

$$p_2 - p_1 + p_0 = -2(p-1),$$

where p_ν is the number of ν-simplices in the triangulation. For example, the sphere can be triangulated by a tetrahedron; here $p_2 = 4$, $p_1 = 6$, $p_0 = 4$, so $2(p-1) = -2$ and $p = 0$.

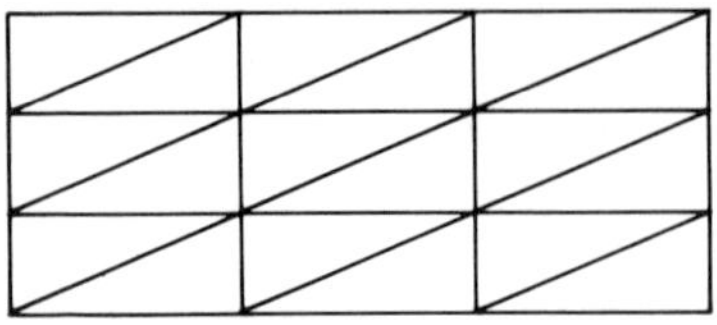

For the torus we have the triangulation above, with opposite sides identified, so $p_2 = 18$, $p_1 = 27$, $p_0 = 9$, hence $2(p-1) = 0$, $p = 1$.

To obtain a formula for p in terms of the Riemann surface we take a triangulation of the sphere which includes all the branch points as vertices and repeat this triangulation on the n sheets covering the sphere. For the sphere we have $p_2 - p_1 + p_0 = 2$, hence in each sheet we have

$$p_2^{(\nu)} - p_1^{(\nu)} + p_0^{(\nu)} = 2 \quad (\nu = 1, \ldots, n).$$

For the triangulation of the whole surface, $p_2 = \Sigma\, p_2^{(\nu)}$, $p_1 = \Sigma\, p_1^{(\nu)}$, but $p_0 = \Sigma\, p_0^{(\nu)} - w$, where $w = \Sigma\, w_i$ and w_i is the ramification index at the ith branch point, because n sheets come together at a branch point. Thus

$$p_2 - p_1 + p_0 = -2(p-1) = 2n - w$$

and it follows that

$$p = \frac{1}{2}w - n + 1. \tag{4}$$

Later (in section 4.6) we shall identify this number with g.

Exercises

1. Show that $\mathbf{R}(x, y)$, where $x^2 + y^2 + 1 = 0$, is not rational, but it becomes so on extending the ground field to $\mathbf{C}$.
2. Show that the equation $x^3 + y^3 = 1$ does not define a rational function field. (*Hint*: Apply Fermat's last theorem for exponent 3.)
3. Show that the equation $x^4 - y^2 = 1$ does not define a rational function field.
4. Show that every function field of genus 0 has a prime divisor of degree 1 or 2. (*Hint*: Use Exercise 2 of section 4.3.)
5. Using the irreducibility of the defining equation, show that the Riemann surface of a function field is connected.
6. Show that any algebraic curve C of degree n, with branch points of multiplicities $r_1, r_2, \ldots$, such that $(n-1)(n-2) = \Sigma\, r_i(r_i - 1)$, is rational. (*Hint*: Consider a curve of degree $n-1$ with each point P_i of multiplicity r_i of C as $(r_i - 1)$-fold point, and $2n-3$ other points on C. Such curves form an r-parameter family, where

$$2r \geqslant (n-1)(n+2) - \sum r_i(r_i - 1) - 2(2n-3) = 2.$$

Show that for $r > 1$ there is a curve B of the family containing two points of C not among the base points. Verify that the number of points common to B and C is greater than $n(n-1)$ and derive a contradiction.)

4.5 DERIVATIONS AND DIFFERENTIALS

One of our tasks will be to replace the inequality in Riemann's theorem by an equation (the Riemann–Roch formula), and this requires the use of differentials. In accordance with our programme we shall introduce the latter in algebraic fashion as linear functionals on the space of derivations.

By a *derivation* on a ring R one understands a mapping $D : R \to R$ which is linear and such that

$$D(xy) = Dx \,.\, y + x \,.\, Dy. \tag{1}$$

Examples are the derivations familiar from elementary analysis. If R is a k-algebra, a derivation of R/k or of R *over* k is a derivation of R which vanishes on k. From (1) it is clear that $D1 = 0$; more generally, any derivation on a field must vanish on the prime subfield. The next result shows that on a function field any derivation is determined by its effect on a separating element.

Proposition 5.1

(i) *Let R be an integral domain with field of fractions K. Then any derivation on R has a unique extension to K.*

(ii) *Let K/E be a separable algebraic extension; any derivation on E has a unique extension to K.*

Proof

(i) Any element u of K has the form $u = a/b$, $a, b \in R$. If a derivation D on R has an extension to K, it must satisfy

$$Da = D(ub) = Du \,.\, b + u \,.\, Db,$$

hence we obtain the usual formula

$$D\left(\frac{a}{b}\right) = \frac{Da}{b} - \frac{a}{b^2} Db, \tag{2}$$

and it is clear that this defines the desired extension. For if $u = a_1/b_1$ is another representation, then $ab_1 = ba_1$, hence

$$a \,.\, Db_1 + Da \,.\, b_1 = Db \,.\, a_1 + b \,.\, Da_1$$

or equivalently,

$$b \,.\, Da_1 - a \,.\, Db_1 = b_1 \,.\, Da - a_1 \,.\, Db.$$

If we divide both sides by bb_1 and remember that $a/b = a_1/b_1$, we obtain

$$\frac{Da_1}{b_1} - \frac{a_1}{b_1^2} \,.\, Db_1 = \frac{Da}{b} - \frac{a}{b^2} Db,$$

which shows that D is well-defined by (2).

(ii) Any element α of K satisfies an irreducible equation over E:

$$f(\alpha) = \alpha^n + c_1\alpha^{n-1} + \ldots + c_n = 0.$$

If D has an extension to K, we must have

$$f^D(\alpha) + f'(\alpha)D\alpha = 0, \tag{3}$$

where f^D is the polynomial with coefficients Dc_i and f' is the usual derivative. Since α is separable over E, $f'(\alpha) \neq 0$, and (3) is a linear equation which determines $D\alpha$ uniquely.■

Now it is clear that for any function field K/k, any separating element x of K defines a unique derivation D_x such that $D_x x = 1$. For on $k[x]$ we have the usual derivative with respect to x, defined by $D_x x^n = nx^{n-1}$, together with linearity. By Proposition 5.1 (i) this extends to $k(x)$ and by (ii) to K. This derivation of K is also denoted by d/dx; thus for any $y \in K$ we write

$$\frac{dy}{dx} = D_x y.$$

Let K/k be a function field. The set D of all derivations of K over k is easily seen to form a vector space over k, with uD defined as the map $x \mapsto uDx$. Let D be any derivation of K/k and x a separating element of K. If $Dx = 0$, then $D = 0$, by what has been said. Otherwise we can form the derivation $C = (Dx)^{-1}D$. Clearly $Cx = 1 = dx/dx$; thus $C - d/dx$ vanishes on x and so must be zero. Hence $du/dx = (Dx)^{-1}Du$, and this proves the familiar:

Chain rule

For any derivation D of a function field K/k and any separating element x of K, we have

$$Du = \frac{du}{dx} Dx. \blacksquare \tag{4}$$

As an illustration let us take any $y \in K$; it will be separable over $k(x)$ and so will have a minimal equation, which can be written as a polynomial in x and y:

$$f(x, y) = 0. \tag{5}$$

For any derivation D of K/k we have

$$f_x Dx + f_y Dy = 0,$$

where f_x, f_y denote the usual partial derivatives of f with respect to x, y respectively. Suppose that $D = d/dx$; then $Dx = 1$, $Dy = dy/dx$, and so we find another familiar rule:

$$\frac{dy}{dx} = -f_x/f_y. \tag{6}$$

For any function field K/k we shall define a *differential* as a linear functional on the space $\mathscr{D}$ of derivations of K/k. For any $u \in K$ let us put du for the linear functional which maps D to Du:

$$\langle D, du\rangle = Du. \tag{7}$$

Since $D(uv) = u \,.\, Dv + Du \,.\, v$, we have $d(uv) = u \,.\, dv + du \,.\, v$. With the help of differentials we can describe the separating elements of a function field. We recall that in a field of prime characteristic p, the mapping $x \mapsto x^p$ is an endomorphism; its image, for K, is denoted by K^p, thus K^p is the set of all pth powers in K.

Proposition 5.2
Let K/k be a function field, where k is perfect of prime characteristic p. Then for any $x \in K$ the following conditions are equivalent:

(a) *x is a separating element for K/k;*
(b) *dx is a K-basis for the space of differentials;*
(c) *$x \notin K^p$.*

Proof
(a) $\Rightarrow$ (*b*). If x is a separating element, any derivation of K/k is determined by its effect on x. As we have seen, if $Dx = u$, then $D = u \,.\, d/dx$, so the space of derivations is one-dimensional with basis d/dx, hence its dual has the basis dx over K.

(b) $\Rightarrow$ (*c*). If $x = y^p$, then $dx = py^{p-1}dy = 0$, so when (c) fails, dx cannot be a basis.

(*c*) $\Rightarrow$ (*a*). By Proposition 3.2.4, K has a separating element y say, and $K = k(y, z)$ for some z (by the theorem of the primitive element). If $[K : k(y)] = n$, then z satisfies an irreducible equation $f(y, z) = 0$ of degree n in z. Hence z^p satisfies

$$f(y, z)^p = g(y^p, z^p) = 0,$$

which is irreducible over $k(y^p)$, for otherwise $f(y, z)$ would also be reducible. Since $g(y^p, z^p)$ is again of degree n in the second argument, we have

$$[K^p : k(y^p)] = [k(y^p, z^p) : k(y^p)] = [K : k(y)] = n. \tag{8}$$

Further, $[k(y):k(y^p)]=p$, hence

$$[K:k(y^p)]=[K:k(y)][k(y):k(y^p)]=np,$$

whereas (8) shows that

$$[K:k(y^p)]=[K:K^p][K^p:k(y^p)]=n[K:K^p].$$

A comparison shows that

$$[K:K^p]=p. \tag{9}$$

We have to show that the extension $K/k(x)$ is separable. If not, let K_1 be the maximal separable extension of $k(x)$ in K. Then $K_1K^p \supseteq K^p(x) \supset K^p$, because $x \notin K^p$. Hence by (9), $K_1K^p = K$. But K/K_1 is purely inseparable, so if $K_1 \neq K$, then $K_1 \subseteq K^p$ and hence $K_1K^p \neq K$, which is a contradiction. Hence $K_1 = K$ and x is indeed a separating element.■

Corollary 5.3
Let K/k be a function field, where k is perfect. Then a uniformizer (at any place) is a separating element for K.

For by Proposition 5.2, if x is not separating, then $x=y^p$ and so x cannot be a uniformizer.■

We shall want to associate a divisor to each differential, and for simplicity we shall assume that k is algebraically closed. At any place $\mathfrak{P}$ of K with valuation $w_{\mathfrak{P}}$ choose a uniformizer t and define

$$w_{\mathfrak{P}}(y\,dx)=w_{\mathfrak{P}}\left(y\frac{dx}{dt}\right). \tag{10}$$

This definition is easily seen to be independent of the choice of t. For if t' is another uniformizer at $\mathfrak{P}$, then $t'=c_1t+c_2t^2+\ldots$, where $c_1\neq 0$, hence $w_{\mathfrak{P}}(dt/dt')=0$ and so

$$w_{\mathfrak{P}}\left(y\frac{dx}{dt'}\right)=w_{\mathfrak{P}}\left(y\frac{dx}{dt'}\cdot\frac{dt'}{dt}\right)=w_{\mathfrak{P}}\left(y\frac{dx}{dt}\right).$$

Moreover, the numbers (10) vanish for almost all $\mathfrak{P}$. For K is a separable extension of $k(x)$, for suitable x, say $K=k(x,z)$, with irreducible equation $f(x,z)=0$. Only finitely many places are branch points for z; at any other place $x-x_0$ can be taken as uniformizer and then $dx/dt=1$. Further, $w_{\mathfrak{P}}(y)=0$ at almost all places and this shows that only finitely many numbers (10) are non-zero. We can therefore define a divisor $(y\,dx)$ as

$$(y\,dx)=\prod \mathfrak{P}^{w_{\mathfrak{P}}(y\,dx)}. \tag{11}$$

The places where $w_{\mathfrak{P}}(y\,dx)$ is positive or negative are again called *zeros* respectively *poles* of $y\,dx$. In characteristic 0 the divisor of a differential can be expressed in terms of the different or ramification divisor. Clearly it is enough to take the case dx:

Proposition 5.4
Let K/k be a function field, where k is algebraically closed of characteristic 0. Then for any $x \in K\backslash k$,

$$(dx) = \mathfrak{d}_{K/k(x)} \cdot (x)_\infty^{-2}, \tag{12}$$

where $\mathfrak{d}_{K/k(x)}$ is the ramification divisor and $(x)_\infty$ is the divisor of poles of x.

Proof
Let $\mathfrak{P}$ be any place of K with ramification index e (over $k(x)$) and choose a uniformizer t. Suppose first that $\mathfrak{P} \nmid (x)_\infty$, thus $\mathfrak{P}$ is not a pole of x, and let x_0 be the value of x at $\mathfrak{P}$. Then

$$x - x_0 = a_1 t^e + a_2 t^{e+1} + \dots, \quad a_1 \neq 0.$$

Hence $dx = (ea_1 t^{e-1} + \dots)dt$ and so the contribution to (12) is t^{e-1}. If $\mathfrak{P}$ is a pole of x, we have

$$x^{-1} = a_1 t^e + a_2 t^{e+1} + \dots,$$

hence

$$x = a_1^{-1} t^{-e} - a_1^{-2} a_2 t^{1-e} + \dots$$

and now $dx = (-ea_1^{-1} t^{-e-1} + \dots)\, dt$, so the contribution is t^{-e-1}. It follows that

$$(dx) = \prod{}' \mathfrak{P}^{e-1} \prod{}'' \mathfrak{P}^{-e-1} = \prod \mathfrak{P}^{e-1} \prod{}'' \mathfrak{P}^{-2e} = \mathfrak{d} \cdot (x)_\infty^{-2},$$

where Π'' is the product over the poles of x and Π' the product over the rest.■

The degree of the ramification divisor,

$$w_x = \sum (e_i - 1),$$

is called the *critical order* of x. Bearing in mind that $d((x)_\infty) = [K : k(x)]$, we obtain the following formula for the degree of a differential divisor:

$$\deg(y\, dx) = w_x - 2[K : k(x)]. \tag{13}$$

If x, y are any separating elements of K, then the formula

$$dy = \frac{dy}{dx}\, dx$$

shows that dx and dy differ only by the function dy/dx and so lie in the same divisor class. This class is denoted by W and is called *canonical divisor class*; it is an important invariant of the field K. Its degree is given by (13); this shows that W is not usually the principal class. Later, in section 4.6, we shall prove the formula

$$d(W) = w_x - 2[K : k(x)] = 2(g-1),$$

where g is the genus; this will show that g is the geometric genus p, obtained in section 4.4.

In the classical theory there is a notion of integration, arising as the inverse of differentiation; we shall use the same principle in defining integrals in what follows. With every function f of x one associates an integral $\int f\,dx$, but strictly speaking the integral is associated with the differential $f\,dx$, not with the function. By Cauchy's theorem any integral around a closed simple contour Γ is equal to $2\pi i$ times the sum of the residues at the poles within Γ. It is this notion for which we shall find an abstract analogue. In what follows we shall restrict ourselves to fields of characteristic 0.

Let K/k be a function field (of characteristic 0) and $\mathfrak{P}$ a place of K, with uniformizer t; given a differential $y\,dx$, we define the *residue* of $y\,dx$ at $\mathfrak{P}$ as

$$\mathrm{res}_{\mathfrak{P}}(y\,dx) = a_{-1} \quad \text{where } y\frac{dx}{dt} = \sum a_i t^i. \tag{14}$$

This definition is independent of the choice of uniformizer. For if t' is another uniformizer, consider any term $t'^r dt'$. Writing $t' = \sum c_\nu t^\nu$, we have for $r \neq -1$,

$$t'^r dt' = \frac{1}{r+1} d(t'^{r+1}) = \frac{1}{r+1} d\left(\left(\sum c_\nu t^\nu\right)^{r+1}\right),$$

and the differential of the series on the right contains no term in t^{-1}. When $r = -1$, we have

$$t'^{-1} dt' = \left(\sum c_\nu t^\nu\right)^{-1}\left(\sum \nu c_\nu t^{\nu-1}\right) dt,$$

and on expanding the right-hand side (bearing in mind that $c_1 \neq 0$, $c_\nu = 0$ for $\nu < 1$), we obtain $t^{-1}dt$ + higher terms. This shows that (14) is unaffected by a change of uniformizer.

In the classical case it is a consequence of Cauchy's theorem that the sum of the residues of a meromorphic function is zero (cf. e.g. Ahlfors (1966)). This still holds in the present context; to prove it we shall first make a reduction to the rational function field.

Lemma 5.5.

Let K/k be a function field, where k is algebraically closed of characteristic 0. Fix $x \in K\backslash k$ and put $E = k(x)$. If $\mathfrak{p}$ is any place of E and $\mathfrak{P}_1, \ldots, \mathfrak{P}_r$ are the places of K above $\mathfrak{p}$, then for any $u \in K$,

$$\sum \mathrm{res}_{\mathfrak{P}_i}(u\,dx) = \mathrm{res}_{\mathfrak{p}}(\mathrm{Tr}_{K/E}(u)dx). \tag{15}$$

Proof

Since k is algebraically closed, we may by a suitable linear transformation or replacing x by x^{-1}, take $\mathfrak{p}$ to be a zero of x. Let t_i be a uniformizer at $\mathfrak{P}_i$ and e_i the ramification index, so that

$$\mathfrak{p} = \mathfrak{P}_1^{e_1} \dots \mathfrak{P}_r^{e_r}.$$

If we replace t_i by $t_i(1 + fx^h)$, where h is a positive integer and f is an element of K integral over $k[x]$, then for large enough h, t_i remains a uniformizer while for suitable choice of f we now have $K = E(t_i)$. At $\mathfrak{P}_1$ we have

$$u\,dx = \sum_{\mu} c_\mu t_1^{\mu-1} dt_1 \quad \text{where } c_\mu \in k,$$

so by linearity it is enough to prove (15) for $t_1^{\mu-1}dt_1$. Writing w_i for the valuation at $\mathfrak{P}_i$, we may choose the t_i so that $w_i(t_j) = \delta_{ij}$, by the approximation theorem. Then it is clear that $\operatorname{res}_{\mathfrak{P}_i}(t_j^{\mu-1}dt_j) = 0$ for $i \neq j$, so we need only show that

$$\operatorname{res}_{\mathfrak{P}_1}(t_1^{\mu-1}dt_1) = \operatorname{res}_{\mathfrak{p}}\left(\operatorname{Tr}_{K/E}\left(t_1^{\mu-1}\frac{dt_1}{dx}\right)dx\right).$$

The minimal polynomial for t_1 over E has the form

$$\prod (t_1 - \lambda_\nu) = t_1^n - a_1 t_1^{n-1} + \dots + (-1)^n a_n.$$

We have $a_n = N_{K/E}(t_1) = \Pi\lambda_\nu$ and $v_{\mathfrak{p}}(a_n) = f_1 = 1$, because k is algebraically closed. Hence

$$\operatorname{Tr}_{K/E}\left(t_1^{-1}\frac{dt_1}{dx}\right) = \sum \lambda_\nu^{-1}\frac{d\lambda_\nu}{dx} = a_n^{-1}\frac{da_n}{dx}.$$

Since $v_p(a_n) = 1$, it follows that $da_n/a_n = (x^{-1} + \dots)dx$, where the dots indicate higher terms, and so

$$\operatorname{Tr}_{K/E}\left(t_1^{-1}\frac{dt_1}{dx}\right) = x^{-1} + \dots .$$

Hence

$$\sum \operatorname{res}_{\mathfrak{P}_i}(t_1^{-1}dt_1) = 1 = \operatorname{res}_{\mathfrak{p}}\left(\operatorname{Tr}_{K/E}\left(t_1^{-1}\frac{dt_1}{dx}\right)dx\right),$$

and this establishes (15) for $u\,dx = t_1^{-1}dt_1$. For general $\mu \neq 0$ we have

$$\operatorname{Tr}_{K/E}\left(t_1^{\mu-1}\frac{dt_1}{dx}\right) = \frac{1}{\mu}\operatorname{Tr}_{K/E}(dt_1^\mu/dx) = \frac{1}{\mu}\sum_{\nu} d\lambda_\nu^\mu/dx = \frac{1}{\mu}\frac{d}{dx}\left(\sum \lambda_\nu^\mu\right).$$

Now the series on the right has no term in $x^{-1}dx$; it follows that the residue is 0, and this proves (15) generally.■

We can now achieve our aim, of showing that the sum of the residues of any differential is zero.

Theorem 5.6

Let K/k be a function field, where k is algebraically closed of characteristic 0. Given $x, y \in K$, the sum of the residues of $y\,dx$ at all places of K is zero.

Proof

By (15) we have

$$\sum \mathrm{res}_{\mathfrak{P}}(y\,dx) = \sum \mathrm{res}_{\mathfrak{p}}(\mathrm{Tr}_{K/E}(y)dx),$$

where $E = k(x)$ and on the right the sum is taken over all places $\mathfrak{p}$ of E/k. Thus we have to show that $\Sigma\, \mathrm{res}_{\mathfrak{p}}(udx) = 0$ for any $u \in k(x)$. We express u as a sum of partial fractions; by linearity it is enough to consider the case $u = (x-a)^{\mu}$. If $\mu \neq -1$, this gives 0 everywhere. For $\mu = -1$ we have

$$\mathrm{res}_{x-a}\left(\frac{dx}{x-a}\right) = 1, \qquad \mathrm{res}_{1/x}\left(\frac{dx}{x-a}\right) = -1,$$

so we again obtain zero and the result follows.■

Any differential on K can be used to define a k-linear functional on the adele ring Λ as follows. Let $y\,dx$ be a differential on K and $u = (u_{\mathfrak{P}})$ an adele. We define

$$\langle u, y\,dx\rangle = \sum \mathrm{res}_{\mathfrak{P}}(u_{\mathfrak{P}}\, y\,dx),$$

where the sum on the right is extended over all places of K. From this point of view Theorem 5.6 just states that $y\,dx$ vanishes on the principal adeles. We note that $y\,dx$ also vanishes on a subspace $\Lambda(\mathfrak{a})$, for some divisor $\mathfrak{a}$. For if $y\,dx$ has its poles confined to $\mathfrak{a}$, then for any $u \in \Lambda(\mathfrak{a})$, $uy\,dx$ has no poles and so $\langle u, y\,dx\rangle = 0$. Thus $y\,dx$ vanishes on $\Lambda(\mathfrak{a}) + K$ and so may be regarded as a linear functional on $\Lambda/\Lambda(\mathfrak{a}) + K$. We shall now show that conversely, every linear functional on $\Lambda/\Lambda(\mathfrak{a}) + K$ is represented by a differential.

Let $\mathfrak{a}$ be any divisor and let ω be a linear functional on Λ vanishing on $\Lambda(\mathfrak{a}) + K$. Further, let $x \in K\backslash k$; we claim that $\omega = y\,dx$ for a suitable $y \in K$. Suppose that dx vanishes on $\Lambda(\mathfrak{b}) + K$ and take an integral divisor $\mathfrak{c}$ to be fixed later. Let us write $D(\mathfrak{c}^{-1})$ for the space of linear functionals on $\Lambda/\Lambda(\mathfrak{c}) + K$. Then

$$u \mapsto udx \text{ is a linear map } L(\mathfrak{b}\mathfrak{c}^{-1}) \to D(\mathfrak{c}^{-1}),$$

$$u \mapsto u\omega \text{ is a linear map } L(\mathfrak{a}\mathfrak{c}^{-1}) \to D(\mathfrak{c}^{-1}).$$

Both maps are k-linear and injective, hence the sum of the dimensions of the images is, by Theorem 3.1:

$$l(\mathfrak{a}\mathfrak{c}^{-1}) + l(\mathfrak{b}\mathfrak{c}^{-1}) \geqslant 2d(\mathfrak{c}) - d(\mathfrak{a}) - d(\mathfrak{b}) + 2 - 2g. \tag{16}$$

On the other hand, $\dim_k D(\mathfrak{c}^{-1}) = \delta(\mathfrak{c}^{-1}) = d(\mathfrak{c}) + g - 1$, and this is less than the right-hand side of (16), for a suitable divisor $\mathfrak{c}$. In this case the images have non-zero intersection; we thus have $u\omega = v\,dx$, hence $\omega = u^{-1}v\,dx$, which is what we wished to show. We state the result as:

Theorem 5.7
For a function field K/k, where k is algebraically closed of characteristic 0, the differentials are precisely the linear functionals on the adele ring which vanish on $\Lambda(\mathfrak{a}) + K$, for some divisor $\mathfrak{a}$.■

This result is the special case, for algebraic functions of one variable, of the Serre duality theorem (for locally free sheaves on a compact complex manifold, cf. Hartshorne (1977)). The theorem allows us to interpret $\delta(\mathfrak{a}^{-1})$ as the dimension of the space of differentials with poles confined to $\mathfrak{a}$; hence a divisor $\mathfrak{a}$ is non-special precisely when there are no differentials with poles confined to $\mathfrak{a}$.

The sums of residues considered above correspond to definite integrals. To find the indefinite integral $\int y\,dx$ we have to find a function u such that $du = y\,dx$. Such a function may not exist within K; it is of course possible to regard $\int y\,dx$ as a new (transcendental) function, but we shall not pursue this line.

Exercises

1. Show that the only fields with no derivation other than 0 are $\mathbf{Q}, \mathbf{F}_p$ and their algebraic extensions.
2. Let K/E be an inseparable extension. Show that there exists a non-zero derivation on K/E.
3. Let K/k be a function field, where k is perfect of characteristic p. Show that for any separating element x of K, $K^p(x) = K$.
4. Let K/k be an algebraic function field of r variables, with a separating transcendence basis $x_1, \ldots, x_r$. Show that the d/dx_i form a K-basis for the derivations of K and the dual basis of differentials is given by $dx_i (i = 1, \ldots, r)$.
5. Let K/k be an algebraic function field of r variables. Show that $x_1, \ldots, x_r$ form a separating transcendence basis if and only if $dx_1, \ldots, dx_r$ is a basis of the differentials.
6. Show that a differential $u\,dz$ may also be defined as an adele whose component at a place $\mathfrak{p}$ with uniformizer t is $u \,.\, dz/dt$. Write down the transformation law for a change of uniformizer and verify that the residue is unchanged.
7. Given a place $\mathfrak{p}$ on a function field of positive genus, show that there exists a differential without poles which does not have a zero at $\mathfrak{p}$. (*Hint*: Assume the contrary and find $x \in K\backslash k$ with poles confined to $\mathfrak{p}$. Now take a differential ω without poles and consider $x^{-r}\omega$ for suitable r.)
8. Extend Proposition 5.4 to fields of prime characteristic. Show that (12) holds if K is tamely ramified at each place over ∞.

4.6 THE RIEMANN–ROCH THEOREM AND ITS CONSEQUENCES

We now return to the inequality of Theorem 3.1 and use the information provided by the canonical divisor class found in section 4.5 to establish an equality, known as the Riemann–Roch theorem. Throughout this section, the ground field k is algebraically closed of characteristic 0.

Let C be a divisor class and $\mathfrak{a}_0, \mathfrak{a}_1, \ldots, \mathfrak{a}_n$ any divisors in C. Then $\mathfrak{a}_i\mathfrak{a}_0^{-1} = (x_i)$ is a principal divisor and the element x_i of K is uniquely determined up to a factor in k. We shall say that $\mathfrak{a}_1, \ldots, \mathfrak{a}_n$ are *linearly dependent* if $x_1, \ldots, x_n$ are linearly dependent over k. Clearly this definition is independent of the choice of x_i in $\mathfrak{a}_i$ or of $\mathfrak{a}_0$. Moreover, if $\mathfrak{a}_1, \ldots, \mathfrak{a}_n$ are any integral divisors in C, then for any $\mathfrak{a}_0$ in C, $\mathfrak{a}_1, \ldots, \mathfrak{a}_n$ are linearly dependent if and only if $x_1, \ldots, x_n$, where $(x_i) = \mathfrak{a}_i\mathfrak{a}_0^{-1}$, are linearly dependent elements of $L(\mathfrak{a}_0^{-1})$. This yields:

Lemma 6.1
Let K/k be a function field. For any divisor class C, and for any divisor $\mathfrak{a}$ in C, the maximum number of linearly independent integral divisors in C is

$$N(C) = l(\mathfrak{a}^{-1}).\blacksquare$$

We can now prove the desired relation for $N(C)$:

Theorem 6.2 (*Riemann–Roch theorem*)
Let K/k be a function field. If C is any divisor class and W is the canonical class, then

$$N(C) = d(C) + 1 - g + N(WC^{-1}), \tag{1}$$

where g is the genus.

This formula (1) is also called the *Riemann–Roch formula.*

Proof
By the lemma, if $\mathfrak{a} \in C$, then $N(C) = l(\mathfrak{a}^{-1})$ is the maximum number of linearly independent integral divisors in C. We have to show that with $\mathfrak{a} \in C$,

$$N(WC^{-1}) = l(\mathfrak{a}^{-1}) + d(\mathfrak{a}^{-1}) + g - 1 = \delta(\mathfrak{a}). \tag{2}$$

Here $\delta(\mathfrak{a})$, the index of specialty, is by Theorems 3.4 and 5.7 the dimension of $D(\mathfrak{a})$, the space of differentials divisible by $\mathfrak{a}$. Write $\delta(\mathfrak{a}) = n$ and let $\omega_1, \ldots, \omega_n$ be a k-basis of $D(\mathfrak{a})$, i.e. a maximal linearly independent set of differentials such that $(\omega_i)\mathfrak{a}^{-1}$ is integral. Any integral divisor $\mathfrak{c}$ in WC^{-1} has the form $\mathfrak{c} = (\omega)\mathfrak{a}^{-1}$ for some differential ω, hence by what has been said, the number of linearly independent integral divisors in WC^{-1} is n, so $N(WC^{-1}) = n = \delta(\mathfrak{a})$, i.e. (2).$\blacksquare$

We remark that the theorem gives a relation between $N(C)$ and $N(WC^{-1})$ in terms of known quantities; it is not a complete determination of $N(C)$. But sometimes it can be used for such a determination. We recall that the principal divisors form a class E, called the *principal divisor class.*

Corollary 6.3

For the principal divisor class E, $N(E) = 1, d(E) = 0$. For the canonical divisor class W, $N(W) = g, d(W) = 2g - 2$.

Proof

It is clear that $d(E) = 0$, and $N(E) = 1$ follows because $L(\mathfrak{e})$ contains only the constants and so is one-dimensional. If we now put $\mathfrak{a} = \mathfrak{e}$ in (2), we obtain

$$N(W) = 1 + 0 + g - 1 = g.$$

Secondly take $\mathfrak{a} \in W$; then (2) becomes $N(E) = l(\mathfrak{a}^{-1}) + d(\mathfrak{a}^{-1}) + g - 1$, i.e. $1 = N(W) - d(W) + g - 1$, which simplifies to $d(W) = 2g - 2$.■

A differential ω is said to be of the *first kind* if it has no poles, i.e. $\omega \equiv 0 \pmod{\mathfrak{e}}$. By (2) we see that the space of such differentials has dimension $N(W) = g$. Thus there are non-zero differentials of the first kind precisely when $g > 0$.

We recall that a divisor class C is called *special* if $N(C) > d(C) + 1 - g$, or equivalently, by (1), if $N(WC^{-1}) > 0$, and *non-special* if equality holds. Thus C is special if and only if it contains a divisor $\mathfrak{a}$ which divides some divisor in the canonical class. For example, by Corollary 6.3, W is always special, while E is special precisely when $g > 0$. The following lemma is useful in proving that certain divisor classes are non-special.

Lemma 6.4

For any divisor class $C \neq E$, if $d(C) \leqslant 0$, then $N(C) = 0$.

For if $N(C) \neq 0$, then $N(C) = l(\mathfrak{a}^{-1}) > 0$ for any divisor $\mathfrak{a}$ in C, hence there is a function with poles confined to $\mathfrak{a}$. It follows that C contains an integral divisor $\mathfrak{a}$, so $d(C) = d(\mathfrak{a}) \geqslant 0$, and here equality is excluded because $\mathfrak{a} \neq \mathfrak{e}$.■

This lemma is essentially a way of saying that all non-constant functions have poles. It leads to the following criterion for a divisor class to be special.

Proposition 6.5

Let C be any divisor class in a function field.

(i) *If $d(C) > 2g - 2$ or $d(C) = 2g - 2$ and $C \neq W$, then C is non-special;*

(ii) *If $d(C) < g - 1$ or $d(C) = g - 1$ and C contains an integral divisor, then C is special.*

Proof

(i) Let us replace C by WC^{-1} in the Riemann–Roch formula:

$$N(WC^{-1}) = d(WC^{-1}) + 1 - g + N(C) = d(W) - d(C) + 1 - g + N(C).$$

Inserting the value $d(W) = 2g - 2$ from Corollary 6.3, we obtain

$$N(WC^{-1}) = N(C) - d(C) + g - 1. \tag{3}$$

Now the hypothesis states that $d(WC^{-1}) < 0$ or $d(WC^{-1}) = 0$ and $WC^{-1} \neq E$, hence by the lemma, $N(WC^{-1}) = 0$ and so $N(C) = d(C) + 1 - g$, by (3), so C is non-special. To prove (ii) we start from the Riemann–Roch formula (1). If $d(C) < g - 1$, (1) shows that $N(C) < N(WC^{-1})$, so $N(WC^{-1}) > 0$ and this means that C is special, by (2). The same holds if $d(C) = g - 1$ and C contains an integral divisor, for then, by Lemma 6.1, $N(WC^{-1}) \geqslant N(C) > 0$.■

Thus in low degrees all divisors are special, while for sufficiently high degrees all are non-special.

Let us return to the Riemann–Roch formula (1). We note that it determines the divisor class W as well as the integer g uniquely. For if W' is any divisor class and g' any integer satisfying

$$N(C) = d(C) + 1 - g' + N(W'C^{-1}) \quad \text{for all } C, \tag{4}$$

let us put C equal to E, W' in turn: we find that $N(W') = g'$, $d(W') = 2g' - 2$, by Corollary 6.3. We now apply Proposition 6.5 with (4) in place of (1) and find that for $d(C) > 2g' - 2$,

$$N(C) = d(C) - g' + 1.$$

Hence for any C such that $d(C) > \max\{2g - 2, 2g' - 2\}$,

$$N(C) = d(C) - g + 1 = d(C) - g' + 1,$$

therefore $g' = g$. It now follows that $d(W') = 2g - 2$ and if $W' \neq W$, then by Proposition 6.5 (i), $N(W') = d(W') + 1 - g$, i.e. $g = 2g - 2 + 1 - g$, whence $1 = 0$. This contradiction shows that $W' = W$.

As an illustration let us consider the simplest case: $K = k(x)$. Take any divisor class C such that $d(C) \geqslant 0$ and let $\mathfrak{a} \in C$. We have

$$l(\mathfrak{a}^{-1}) = d(\mathfrak{a}) + 1,$$

for this formula holds when $\mathfrak{a} = \mathfrak{e}$, and each pole increases l by 1, while each zero decreases it by 1. When $d(C) < 0$, we have $l(\mathfrak{a}^{-1}) = 0$, so that any divisor $\mathfrak{a}$ satisfies

$$l(\mathfrak{a}^{-1}) + d(\mathfrak{a}^{-1}) - 1 = \max\{0, d(\mathfrak{a}^{-1})\}.$$

This shows again that $g = 0$ in this case (cf. Theorem 4.1). A divisor class C on $k(x)$ is special for $d(C) < -1$ and non-special for $d(C) \geqslant -1$. For any function field K of genus 0 this method shows (as in the proof of Theorem 4.1) that $K = k(x)$.

We can now also establish the formula for the genus in terms of the critical order. It is convenient to derive a more general formula for extensions of function fields, which has the desired formula as a special case.

Theorem 6.6
Let E/k be a function field, where k is algebraically closed of characteristic 0 and let K be a finite algebraic extension of E, of degree n. Write g_E, g_K for the genus of E, K respectively and δ for the degree of the different of K/E. Then

$$2(g_K - 1) = 2n(g_E - 1) + \delta. \tag{5}$$

Proof
Let ω be a non-zero differential of E; this may be written as $\omega = y\,dx$, where $x, y \in E$; its degree is $2(g_E - 1)$, by Corollary 6.3. We now regard x, y as elements of K and compute its degree in K. Let $\mathfrak{P}$ be any place in K with uniformizer t and suppose that $\mathfrak{P} \mid \mathfrak{p}$, where $\mathfrak{p}$ is a place of E, with uniformizer z.

We have $z = ut^e$, where e is the ramification index and u is a unit at $\mathfrak{P}$. Now $dz = t^e du + eut^{e-1}dt$, hence for any $r \in \mathbf{Z}$,

$$z^r dz = u^r t^{e(r+1)} du + eu^{r+1} t^{e(r+1)-1} dt.$$

On the left the $\mathfrak{p}$-value is r, while on the right the $\mathfrak{P}$-value is

$$e(r + 1) - 1 = er + e - 1.$$

Writing $v_\mathfrak{p}$, $w_\mathfrak{P}$ for the normalized valuations, we have

$$w_\mathfrak{P}(y\,dx) = e_\mathfrak{P} v_\mathfrak{p}(y\,dx) + e_\mathfrak{P} - 1.$$

If we sum over all places and recall from Proposition 1.3 that $\delta = \Sigma(e_\mathfrak{P} - 1)$, while of course $\Sigma e_\mathfrak{P} = n$, we obtain the degree (in K) of $y\,dx$ on the left, which we have seen is $2(g_K - 1)$, while on the right we have $2n(g_E - 1) + \delta$. This establishes (5).■

Sometimes (5) is expressed in the form of a relative genus formula (Hurwitz formula):

$$g_K - ng_E = \frac{1}{2}\delta - n + 1. \tag{6}$$

In particular, taking $E = k(x)$, we have $g_E = 0$ and so obtain:

Corollary 6.7
For any function field K/k (k algebraically closed of characteristic 0), the genus g is given by

$$g = \frac{1}{2}\delta - (n - 1), \tag{7}$$

where n is the degree of K over a rational subfield $k(x)$ and δ is the degree of its different.■

If we bear in mind that δ is also the degree of the ramification divisor, this formula shows g to be identical with the geometric genus, as defined in section 4.4 (cf. (4) of section 4.4).

As an application let us prove:

Theorem 6.8 (*Lüroth's theorem*)
Any subextension of a pure transcendental extension of one variable is again pure.

Proof
Here we shall assume that the ground field k is algebraically closed of characteristic 0, though the result holds without these restrictions (cf. e.g. A.3, Theorem 5.2.4). Let $K = k(t) \supseteq E \supset k$. Then $g_K = 0$ in (6), hence

$$2n(1 - g_E) = \delta + 2.$$

Since the right-hand side is positive, it follows that $g < 1$, and so $g = 0$. Therefore E is a rational function field, by Theorem 4.1.■

For any function field K the following question is of great importance: what is the least degree of the divisor of poles of an element of K? We recall from Theorem 2.1 that this degree is also $[K : k(x)]$, so there is a function with a simple pole precisely for the rational function field and in no other case. Of course we know from Corollary 3.2 that there is always a function whose divisor of poles has degree $g + 1$. As a first step we prove a theorem showing that in any increasing divisor sequence $(\mathfrak{a}_n)$ with $d(\mathfrak{a}_n) = n$ there are precisely g 'gaps'.

Theorem 6.9 (*Max Noether*)
Let K/k be a function field, where k is algebraically closed of characteristic 0. Given any sequence of places $\mathfrak{P}_1, \mathfrak{P}_2, \ldots$ of K, not necessarily distinct, write $\mathfrak{a}_n = \mathfrak{P}_1 \ldots \mathfrak{P}_n$. Then there are precisely g values of ν for which

$$L(\mathfrak{a}_\nu^{-1}) = L(\mathfrak{a}_{\nu-1}^{-1}).$$

Moreover, these values all occur for $\nu \leqslant 2g - 1$.

These g values of ν are called exceptional values or *gaps* in the sequence $\mathfrak{P}_1, \mathfrak{P}_2, \ldots$; the theorem just tells us that there is no function in K with divisor of poles exactly equal to $\mathfrak{a}_\nu$.

Proof
Write $\lambda_\nu = l(\mathfrak{a}_\nu^{-1}) + d(\mathfrak{a}_\nu^{-1})$; we note that when $\mathfrak{a}_\nu^{-1}$ is non-special, then $\lambda_\nu = 1 - g$, while for special a_ν^{-1}, $\lambda_\nu > 1 - g$. Now $\lambda_0 = 1 + 0 = 1$, while for $\nu > 2g - 2$, $\lambda_\nu = 1 - g$ by Proposition 6.5. Further,

$$\lambda_\nu = \begin{cases} \lambda_{\nu-1} & \text{if there is no gap at } \nu, \\ \lambda_{\nu-1} - 1 & \text{if there is a gap at } \nu. \end{cases}$$

Thus λ_ν is a decreasing integer-valued function varying from 1 to $1-g$ by steps of 1; it follows that there are exactly g gaps in the sequence $\mathfrak{a}_0, \ldots, \mathfrak{a}_{2g-1}$.■

This theorem does not tell us exactly where the gaps are, merely that they occur among the first $2g-1$ terms in the series. For $g=0$, the case of a rational function field, there are no gaps, as we already know from Theorem 4.1, while for $g=1$ there is just one gap. In the case where the divisor sequence consists of the powers of a single place: $\mathfrak{P}, \mathfrak{P}^2, \ldots$, the gap must be at $\mathfrak{P}$, for if there were a function x with $\mathfrak{P}$ as divisor of poles, then x^r would have $\mathfrak{P}^r$ as divisor of poles.

As a further illustration let us consider the case where for some divisor sequence there is no gap at $\nu=2$; this means that there is a function x whose divisor of poles has degree 2 and so $[K:k(x)]=2$. A function field with this property is said to be *hyperelliptic*. We then have $K=k(x,y)$ for any $y \in K \backslash k(x)$ and y satisfies a quadratic equation over $k(x)$, say $y^2+ay+b=0$, where $a, b \in k(x)$. Here we may take a, b to be polynomials in x, by replacing y by cy, for a common denominator c of a, b, and we may further assume that $a=0$, by replacing y by $y+a/2$. We thus obtain an equation

$$y^2 = f(x), \tag{8}$$

where f is a polynomial in x. Clearly f may be taken to have distinct zeros, for if f had a repeated factor: $f=gh^2$, we could replace y by y/h and so simplify f. Suppose that the degree of f is even: $\deg f = 2m$. On replacing x by $x-\alpha$, where $f(\alpha)=0$, we can arrange that $f(0)=0$, so that x, y have a common zero. If we now put $x' = x^{-1}$, $y' = y/x^2$, then the relation

$$y^2 = x^{2m} + a_1 x^{2m-1} + \ldots + a_{2m-1} x$$

becomes

$$y'^2 = a_{2m-1} x'^{2m-1} + \ldots + a_1 x' + 1.$$

Thus for a given hyperelliptic field the polynomial f in the defining equation (8) can be taken either of even degree $2m$ or of degree $2m-1$.

In factorized form we may write (8) as

$$y^2 = (x-\alpha_1) \ldots (x-\alpha_{2m});$$

as we have seen, taking the degree to be even is no restriction. Each value α_i corresponds to a place $\mathfrak{P}_i$, where $\mathfrak{P}_i^2$ divides $(x-\alpha_i)$ while $\mathfrak{P}_1 \ldots \mathfrak{P}_{2m}$ is the divisor of zeros of y. The different of $K/k(x)$ has degree $\Sigma_1^{2m}(e_i-1)=2m$, while the degree is 2, so by (7) we obtain $g=m-2+1=m-1$. Hence a hyperelliptic field defined by (8), where f has degree $2m$ or $2m-1$, has genus $m-1$. There are $2m$ branch points (when the function f in (8) has odd degree $2m-1$, the remaining branch point is at infinity).

Let us consider the special case of Theorem 6.9 when the places $\mathfrak{P}_i$ all coincide. The gaps are now g integers $\rho_1, \ldots, \rho_g$ for which there is no function with a single pole at a place $\mathfrak{P}$, of order exactly ρ_ν. Moreover, if $\rho_1 > 1$, then we have a function x with a simple pole at $\mathfrak{P}$; hence x^n has a pole of order n at $\mathfrak{P}$ and so there are no gaps, i.e. $g = 0$. We thus have:

Theorem 6.10 (*Weierstrass gap theorem*)
Let K/k be a function field of genus g and $\mathfrak{P}$ any place of K. Then there are g integers

$$0 < \rho_1 < \ldots < \rho_g < 2g, \tag{9}$$

where $\rho_1 = 1$ if $g \neq 0$, such that there is a function with an r-fold pole at $\mathfrak{P}$ if and only if $r \neq \rho_\nu$ $(\nu = 1, \ldots, g)$. ■

Historically Theorem 6.10 came first and Theorem 6.9 was then obtained as a generalization by Max Noether in 1884. Theorem 6.9 also leads to another proof of Corollary 3.2. Any integral divisor $\mathfrak{a}$ of degree $g+1$ can be built up one place at a time: $\mathfrak{a}_0 = \mathfrak{e}, \mathfrak{a}_1, \ldots, \mathfrak{a}_{g+1} = \mathfrak{a}$, and this series has at most g gaps, so $L(\mathfrak{a}^{-1})$ contains a function other than a constant.

We shall find that in general (more precisely for almost all places) the gaps all come in a single block at the beginning, i.e. $\rho_\nu = \nu$ in Theorem 6.10. A place for which this is not so is called a *Weierstrass point*; these points are of great interest because they are invariants of the function field K, independent of its representation. Clearly for $g = 1$ there are no Weierstrass points, because the one and only gap must come at $\nu = 1$. We shall now show that for $g > 1$ there are always Weierstrass points, but only finitely many.

Theorem 6.11
Let K/k be a function field of genus g, where k is algebraically closed of characteristic 0. If K is elliptic, it has no Weierstrass points. A hyperelliptic field K has $2g+2$ Weierstrass points, each with gaps in the positions $1, 3, 5, \ldots, 2g-1$. If K is not hyperelliptic, then $g \geqslant 3$ and there are h Weierstrass points, where

$$2(g+1) < h \leqslant (g-1)g(g+1). \tag{10}$$

Proof
We may suppose that $g > 0$ (since no gaps occur for $g = 0$). It follows that for any place $\mathfrak{P}$ the integers in (9) are such that $\rho_1 = 1$. When $g = 1$, the elliptic case, there are no Weierstrass points, as we have seen. When $g = 2$, there are two gaps and $\rho_1 = 1$, so at any Weierstrass point $\rho_2 > 2$; this means that there is a function with a single pole, of order 2, and as we have seen, K is then

hyperelliptic. This holds provided there is at least one Weierstrass point (which is so, as this proof will show).

Consider the differentials of the first kind; they form a g-dimensional k-space, by Corollary 6.3, and we can choose a basis $\omega_1, \ldots, \omega_g$ so that at a given place $\mathfrak{P}$, ω_ν has a zero of order μ_ν, where

$$0 \leqslant \mu_1 < \mu_2 < \ldots < \mu_g. \tag{11}$$

If f is a function with a single pole at $\mathfrak{P}$, of order ρ, then $f\omega_\nu$ is a differential with a pole of order $\rho - \mu_\nu$ at $\mathfrak{P}$. By Theorem 5.6 the residue at $\mathfrak{P}$ must be 0, so we must have $\rho \neq \mu_\nu + 1$. Thus a gap occurs at $\mu_\nu + 1$ and the exponents ρ_ν for $\mathfrak{P}$ are

$$\rho_\nu = \mu_\nu + 1, \quad \nu = 1, \ldots, g. \tag{12}$$

Since $\rho_1 = 1$, we have $\mu_1 = 0$; in other words, for any place $\mathfrak{P}$ there is a differential of the first kind not vanishing at $\mathfrak{P}$.

If the place $\mathfrak{P}$ is not a Weierstrass point, we have $\rho_\nu = \nu$ and so $\mu_\nu = \nu - 1$. Our aim is to show that this holds for almost all places. Let us fix a place $\mathfrak{P}$ and take a uniformizer t at $\mathfrak{P}$. Then we can express the differentials ω_ν as

$$\omega_\nu = u_\nu(t)dt, \quad \nu = 1, \ldots, g.$$

The corresponding divisors lie in the class W. Now formally we can take the product of r differentials and obtain a divisor in W^r; likewise we can take linear combinations of such products (for the same r) and still get a divisor in W^r. Consider the Wronskian determinant of the ω_ν:

$$\Delta(t)dt^{g(g+1)/2} = \begin{vmatrix} u_1 & u_2 & \ldots & u_g \\ u'_1 & u'_2 & \ldots & u'_g \\ \ldots & \ldots & \ldots & \\ u_1^{(g-1)} & u_2^{(g-1)} & \ldots & u_g^{(g-1)} \end{vmatrix} dt^{g(g+1)/2}$$

At the place $\mathfrak{P}$ the exponent for u_ν begins with t^{μ_ν} and we can adjust the constant factor so that we have

$$u_\nu(t) = t^{\mu_\nu} + \ldots, \quad \nu = 1, \ldots, g.$$

If we insert these expansions in $\Delta(t)$, we find that the first term in the expansion is the determinant

$$\begin{vmatrix} t^{\mu_1} & t^{\mu_2} & \ldots & t^{\mu_g} \\ \mu_1 t^{\mu_1 - 1} & \mu_2 t^{\mu_2 - 1} & \ldots & \mu_g t^{\mu_g - 1} \\ \ldots & \ldots & \ldots & \ldots \end{vmatrix}$$

To evaluate it, we multiply the rows by $1, t, t^2, \ldots, t^{g-1}$ in turn and then divide the columns by $t^{\mu_1}, t^{\mu_2}, \ldots, t^{\mu_g}$. We obtain

$$\begin{vmatrix} 1 & 1 & \dots & 1 \\ \mu_1 & \mu_2 & \dots & \mu_g \\ \mu_1(\mu_1 - 1) & \mu_2(\mu_2 - 1) & \dots & \mu_g(\mu_g - 1) \\ \dots & \dots & \dots & \dots \end{vmatrix}$$

By row operations this is reduced to a Vandermonde determinant, whose value is $\Pi_{i>j}(\mu_i - \mu_j)$ (A.1, p. 192). This is multiplied by a power of t, with exponent

$$\sum \mu_\nu - \sum (\nu - 1) = \sum \mu_\nu - g(g-1)/2.$$

Let us write $\mu = \Sigma\mu_\nu$. By (11), $\mu_\nu \geqslant \nu - 1$, with equality for all ν precisely when $\mathfrak{P}$ is a Weierstrass point; hence $\mu \geqslant g(g-1)/2$, with equality holding precisely at the Weierstrass points. We thus have

$$\Delta(t)dt^{g(g+1)/2} = \left[\prod_{i>j}(\mu_i - \mu_j)t^{\mu - g(g-1)/2} + \dots\right] dt^{g(g+1)/2}, \tag{13}$$

a product of differentials which has no pole, and which has a zero at $\mathfrak{P}$ precisely when $\mathfrak{P}$ is a Weierstrass point. Now the number of zeros of a differential of the first kind is $d(W) = 2(g-1)$, by Corollary 6.3, hence the number of zeros of $\Delta(t)$ is

$$d(W^{g(g+1)/2}) = (g-1)g(g+1).$$

This provides an upper bound for the number of Weierstrass points. More precisely, if the Weierstrass points are $\mathfrak{P}_1, \dots, \mathfrak{P}_h$, with multiplicities $m_1, \dots, m_h$, then

$$\sum_1^h m_i = (g-1)g(g+1), \tag{14}$$

and at a place $\mathfrak{P}_i$ with multiplicity m_i we have, by (13) and (12),

$$m_i = \sum \mu_\nu - g(g-1)/2 = \sum \rho_\nu - g(g+1)/2. \tag{15}$$

We claim that $m_i \leqslant g(g-1)/2$, with equality if and only if $g \leqslant 2$ (the elliptic and hyperelliptic cases).

If x, y are functions whose only pole is $\mathfrak{P}$ and the orders are r, s respectively, then xy has as its only pole $\mathfrak{P}$ of order $r + s$. Thus for any place $\mathfrak{P}$, the set of positive numbers different from $\rho_1, \dots, \rho_g$ is closed under addition. Let r be the least integer occurring as order of poles; then $r > 1$ and the gaps occur at the orders

$$\begin{array}{ccccc} 1 & 2 & 3 & \dots & r-1 \\ r+1 & r+2 & r+3 & \dots & r+r-1 \\ \dots & \dots & \dots & \dots & \dots \\ \lambda_1 r+1 & \lambda_2 r+2 & \lambda_3 r+3 & \dots & \lambda_{r-1}r+r-1 \end{array} \tag{16}$$

For all multiples of r must occur as orders, and whenever $\lambda r+s$ occurs as order, so does $(\lambda+1)r+s, (\lambda+2)r+s, \dots$, so there are no gaps apart from (16). The total number of gaps is g, hence

$$\sum_1^{r-1}(\lambda_i+1)=g, \tag{17}$$

and there are no gaps beyond $2g-1$, whence

$$\lambda_i r+i \leqslant 2g-1, \quad i=1,\dots,r-1. \tag{18}$$

When $r=2$, (17) reads $\lambda_1+1=g$, and then equality holds in (18); clearly this is the only case.

Now the sum of the orders in (16) is

$$\begin{aligned} \sum \rho_\nu &= \sum_{i=1}^{r-1}\sum_{\lambda=0}^{\lambda_i}(\lambda r+i) \\ &= \sum_{i=1}^{r-1}\left(\frac{\lambda_i(\lambda_i+1)}{2}r+i(\lambda_i+1)\right) \\ &= \sum_{i=1}^{r-1}\left[\frac{1}{2r}(\lambda_i r+i)^2-\frac{i^2}{2r}+\frac{\lambda_i r}{2}+i\right] \end{aligned}$$

By (18) we have

$$\begin{aligned} \frac{1}{2r}(\lambda_i r+i)^2-\frac{i^2}{2r}+\frac{\lambda_i r}{2}+i &\leqslant \frac{(2g-1)}{2r}(\lambda_i r+i)-\frac{i^2}{2r}+\frac{\lambda_i r}{2}+i \\ &= \lambda_i\left(\frac{2g-1+r}{2}\right)+i\left(\frac{2g-1}{2r}+1\right)-\frac{i^2}{2r}. \end{aligned}$$

If we sum this expression from $i=1$ to $r-1$ and recall the elementary formulae

$$\sum_1^{r-1} i=r(r-1)/2, \qquad \sum_1^{r-1} i^2=r(r-1)(2r-1)/6,$$

while from (17) $\Sigma\lambda_i=g-r+1$, we obtain

$$\begin{aligned} &\frac{(2g-1+r)(g-r+1)}{2}+\frac{r(r-1)(2g-1+2r)}{4r}-\frac{r(r-1)(2r-1)}{12r} \\ &\qquad = g^2-\frac{(r-1)(r-2)}{6}. \end{aligned}$$

Thus we finally have

$$\sum \rho_v \leqslant g^2 - \frac{(r-1)(r-2)}{6}, \tag{19}$$

where equality holds precisely for $r = 2$, since this is the only case of equality in (18). Going back to (15) we thus find that the multiplicity m_i of the ith Weierstrass point $\mathfrak{P}_i$ satisfies

$$m_i \leqslant \frac{g(g-1)}{2}, \tag{20}$$

with equality if and only if $r = 2$. Consider first the case $r = 2$, when there is a function x with a single pole, of order 2, and so K is a hyperelliptic field. Then $K = k(x, y)$, where y is quadratic over $k(x)$, and as we have seen, there are $2m = 2(g+1)$ branch points $\mathfrak{P}_1, \ldots, \mathfrak{P}_{2m}$. Suppose that at $\mathfrak{P}_i$, $y = y_i$; then the function $(y - y_i)^{-1}$ has a single pole at $\mathfrak{P}_i$ of order 2, hence each of these points is a Weierstrass point for $g > 1$. At any Weierstrass point there is a function with a single pole, of order 2, so this must be a ramification point, i.e. one of the $\mathfrak{P}_i$. Now by (20) (with equality, because now $r = 2$), we have

$$\sum_{i=1}^{2g+2} m_i = (g+1)g(g-1),$$

which we already know from (14). At any of these points the gaps are at orders $1, 3, 5, \ldots, 2g-1$ (by the scheme (16)) and $\Sigma_1^g(2i-1) = g^2$, so equality holds in (19). This proves the assertions in the hyperelliptic case. We have seen that this case must occur when $r = 2$, while $r = 1$ is the elliptic case, so we may now assume that $r \geqslant 3$. For the number of Weierstrass points we have by (14), $h \leqslant (g-1)g(g+1)$; using the bound (20) for m_i in (14), where the inequality is now strict, we find

$$h\frac{g(g-1)}{2} > (g-1)g(g+1),$$

hence $h > 2(g+1)$, and so (10) is proved.■

The lower bound in (10) was established by Hurwitz in 1893.

Exercises

1. Show that any divisor class C contains an integral divisor if and only if $N(C) \geqslant 1$.
2. Prove the Riemann–Roch theorem for the case of genus zero by showing:
   (i) $N(C) = \max\{\alpha + 1, 0\}$, where α ranges over all integers such that $\infty^\alpha \in C$;
   (ii) $N(WC^{-1}) = \max\{-(\alpha+1), 0\}$.
3. Show that $N(W^n) - N(W^{-(n-1)}) = (2n-1)(g-1)$.

4. Show that for each place $\mathfrak{P}$ there is a differential which has as its only pole $\mathfrak{P}$, of order 2, with residue 0.
5. Show that for any places $\mathfrak{P}_1, \mathfrak{P}_2$ there is a differential which has $\mathfrak{P}_1, \mathfrak{P}_2$ as its only poles.
6. Show that for any place $\mathfrak{P}$ and for any principal part of degree n of a differential with zero residue there exists a differential with $\mathfrak{P}$ as its only pole and with the given principal part. (*Hint*: Verify that $N(W\mathfrak{P}^n)$ increases by 1 as n increases by 1.)
7. Show that for any places $\mathfrak{P}_1, \ldots, \mathfrak{P}_n$ with given principal parts whose residues have sum zero there exists a differential with these principal parts.
8. Show that $2N(C) - d(C)$ is invariant under passage to the complementary class $C \mapsto WC^{-1}$. Show that for $d(C) \neq g - 1$ there is no linear combination independent of the given one with the same property; what happens in the exceptional case?
9. With the notation of the proof of Theorem 6.11 show that

$$\lambda_i + \lambda_j \geqslant \begin{cases} \lambda_{i+j} - 1 & \text{if } i + j < r, \\ \lambda_{i+j-r} - 2 & \text{if } i + j \geqslant r. \end{cases}$$

10. Use the Riemann–Roch theorem to show that for a gap at $\mathfrak{P}^r$ there is no differential of the first kind with an $(r-1)$-fold zero at $\mathfrak{P}$.

4.7 ELLIPTIC FUNCTION FIELDS

We shall now consider function fields of genus 1, or elliptic function fields, as they were called in section 4.4, in greater detail. In this case $N(W) = N(E) = 1$, $d(W) = d(E) = 0$ by Corollary 6.3; hence, by the remark after Proposition 6.5, $W = E$, thus the canonical divisor class and the principal class coincide. By Proposition 6.5, a divisor class C is non-special if and only if $d(C) > 0$ or $d(C) = 0$ and $C \neq E$. We shall in this particular case give a separate proof that

$$J \cong \mathbf{C}/\Omega \quad \text{(Abel–Jacobi theorem)},$$

where J is the group of divisor classes of degree zero and Ω is the period lattice (to be defined). For the moment we shall take k algebraically closed and of characteristic 0, and show how to get a bijection between J and the set of prime divisors (places); later the ground field will be taken to be $\mathbf{C}$.

Let K be an elliptic function field and let us fix a place $\mathfrak{o}$. Since the ground field k is algebraically closed, $d(\mathfrak{o}) = 1$, $[\mathfrak{o}]$ is non-special and $l(\mathfrak{o}^{-2}) = 2$, so there is a non-constant function x with poles in $\mathfrak{o}^2$:

$$(x) = \frac{(x)_0}{\mathfrak{o}^2}.$$

Since K/k is not rational, it has no function with pole $\mathfrak{o}$ and so $\mathfrak{o}^2$ is the exact divisor of poles of x; hence $[K:k(x)]=2$. Next $l(\mathfrak{o}^{-3})=3$, so there exists $y\in L(\mathfrak{o}^{-3})\backslash L(\mathfrak{o}^{-2})$, say

$$(y)=\frac{(y)_0}{\mathfrak{o}^3}.$$

Since $L(\mathfrak{o}^{-6})$ has dimension 6 and contains $1, x, x^2, x^3, y, y^2, xy$, there must be a linear relation between them, and here y^2 must occur, for otherwise we would have $K=k(x)$. Thus our relation takes the form

$$y^2+(ax+b)y+cx^3+dx^2+ex+f=0, \quad a, b, \ldots, f\in k.$$

By completing the square we get rid of the terms in y. On comparing poles at $\mathfrak{o}^6$ we see that $c\neq 0$, and we can reduce c to -4 by making the change $y\mapsto 4c^2y$, $x\mapsto -4cx$. Next we replace x by $x+d/12$ to get rid of the term in x^2. There remains the equation

$$y^2=4x^3-g_2x-g_3, \quad \text{where } g_2, g_3\in k. \tag{1}$$

This is called the *Weierstrass normal form* of the equation defining an elliptic function field. If the polynomial on the right of (1) had a repeated zero, say $y^2=4(x-\alpha)^2(x-\beta)$, then K would be the field of rational functions in $y/(x-\alpha)$, which is not the case. Hence this polynomial has distinct zeros, and so its discriminant is non-zero; up to a factor 16 this is $g_2^3-27g_3^2$.

By construction, $\mathfrak{o}$ is the only pole of x or y, so the other places correspond to solutions (x_1, y_1) of (1) in k. We claim that together with $\mathfrak{o}$ this includes all places. For if $\mathfrak{p}$ is a place $\neq\mathfrak{o}$ and I is the ring of integers at $\mathfrak{p}$, then by taking residue classes we have a map $I\to k$. If $x\mapsto x_1$, $y\mapsto y_1$, then (x_1, y_1) is a solution of equation (1) in k, and all solutions clearly occur in this way.

We ask: how are the places distributed over the divisor classes? The answer is given by:

Theorem 7.1

In an elliptic function field K/k there is a bijection between prime divisors and divisor classes of degree 0, given by $\mathfrak{p}\leftrightarrow[\mathfrak{p}/\mathfrak{o}]$, where $\mathfrak{o}$ is an arbitrary but fixed prime divisor.

Proof

Fix $\mathfrak{o}$; then for any prime divisor $\mathfrak{p}$, $[\mathfrak{p}/\mathfrak{o}]$ is a divisor class of degree 0. Conversely, if P is a divisor class of degree 0, then the Riemann–Roch formula states that

$$N(P\mathfrak{o})=d(P\mathfrak{o})+N(P^{-1}\mathfrak{o}^{-1}),$$

and since $P\mathfrak{o}$ is non-special, the right-hand side is 1, so it contains an integral divisor $\mathfrak{p}$, which must be prime because its degree is 1. Thus the mapping is surjective. To see that it is injective, suppose that $[\mathfrak{p}/\mathfrak{o}]=[\mathfrak{q}/\mathfrak{o}]$; then

$[\mathfrak{q}/\mathfrak{p}] = E$, hence if $\mathfrak{p} \neq \mathfrak{q}$, we have $\mathfrak{q}/\mathfrak{p} = (z)$ for some $z \in K$, but then $K = k(z)$, which is a contradiction.■

Theorem 7.1 describes a bijection between J, the group of divisor classes of degree 0, and the points of the Riemann surface R of the curve (1). We shall use this bijection to define an abelian group structure on R. Thus we write

$$\mathfrak{p}_1 + \mathfrak{p}_2 = \mathfrak{p}_3 \quad \text{whenever } \frac{\mathfrak{p}_1}{\mathfrak{o}} \cdot \frac{\mathfrak{p}_2}{\mathfrak{o}} \sim \frac{\mathfrak{p}_3}{\mathfrak{o}}, \quad \text{i.e. } \mathfrak{p}_1\mathfrak{p}_2 \sim \mathfrak{p}_3\mathfrak{o}, \tag{2}$$

where $\sim$ indicates equivalence of divisors. In this way R becomes an abelian group with $\mathfrak{o}$ as neutral element. Let us denote the residues of x, y at $\mathfrak{p} \neq \mathfrak{o}$ by $x\mathfrak{p}, y\mathfrak{p}$ (recall that x, y are finite at each $\mathfrak{p} \neq \mathfrak{o}$). Then we have:

Theorem 7.2 (*Addition theorem*)
Let $\mathfrak{o}, \mathfrak{p}_1, \mathfrak{p}_2, \mathfrak{p}_3$ be all different prime divisors of an elliptic function field such that in the addition just defined,

$$\mathfrak{p}_1 + \mathfrak{p}_2 + \mathfrak{p}_3 = \mathfrak{o}. \tag{3}$$

Then

$$\begin{vmatrix} 1 & x\mathfrak{p}_1 & y\mathfrak{p}_1 \\ 1 & x\mathfrak{p}_2 & y\mathfrak{p}_2 \\ 1 & x\mathfrak{p}_3 & y\mathfrak{p}_3 \end{vmatrix} = 0. \tag{4}$$

If some of the $\mathfrak{p}_i, \mathfrak{o}$ are equal, this still holds with obvious modifications (obtained by differentiating with respect to $\mathfrak{p}_i$). For example if $\mathfrak{p}_1 = -\mathfrak{p}_2 \neq \mathfrak{o}$, then

$$\begin{vmatrix} 1 & x\mathfrak{p}_1 \\ 1 & x\mathfrak{p}_2 \end{vmatrix} = 0.$$

Proof
Equation (3) states that $\mathfrak{p}_1\mathfrak{p}_2\mathfrak{p}_3\mathfrak{o}^{-3} = (z)$, where $z \in L(\mathfrak{o}^{-3})$. Hence $z = \alpha x + \beta y + \gamma$, and here $\beta \neq 0$, because the divisor of poles of z is $\mathfrak{o}^3$. Therefore in (4) we can replace the third column by $z\mathfrak{p}_i$ without changing the value of the determinant. Since $z\mathfrak{p}_i = 0$ for $i = 1, 2, 3$, the result follows. The other cases are proved similarly.■

The result may be interpreted geometrically as follows: (1) represents a plane cubic curve while (4) states that the three points $(x\mathfrak{p}_i, y\mathfrak{p}_i)$ are collinear. So the addition on the cubic is defined as follows: given two points P_1, P_2 on the curve, if their join meets the curve again at P_3, then $P_3 = -(P_1 + P_2)$.

Since an elliptic function field has no functions with simple poles, there is no linear series consisting of single points. There is a series consisting of point pairs $\mathfrak{p}, \mathfrak{q}$ such that $\mathfrak{pq}$ lies in a fixed divisor class, as we see from (2).

We next consider automorphisms of a function field. Let K/k be any

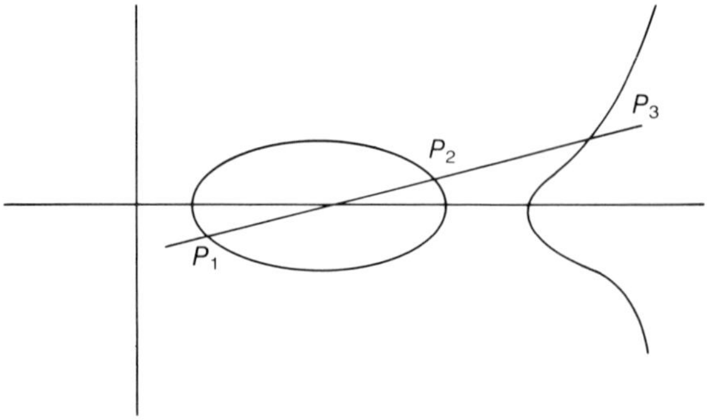

function field and σ any automorphism of K/k, i.e. an automorphism of K leaving k elementwise fixed. Any place $\mathfrak{p}$ of K is described by a certain valuation ring in K and its maximal ideal; by applying σ we obtain another place $\mathfrak{p}^\sigma$. We claim that σ is completely determined by its effect on the places. For if σ is such that $\mathfrak{p}^\sigma = \mathfrak{p}$ for all $\mathfrak{p}$, then any $x \in K\backslash k$ has the same poles and zeros as x^σ and so satisfies $x^\sigma = c_x x$ where $c_x \in k$. Moreover, $(x+1)^\sigma = x^\sigma + 1$, hence

$$c_{x+1}(x+1) = c_x x + 1, \quad \text{and so} \quad c_x = c_{x+1} = 1.$$

This proves:

Proposition 7.3
Any automorphism of a function field K/k is completely determined by the permutation of the places induced by it. ■

In order to find specific automorphisms of an elliptic function field K, let us take two places $\mathfrak{o}, \mathfrak{o}'$ of K, not necessarily distinct. We have $l((\mathfrak{o}\mathfrak{o}')^{-1}) = 2$, so there exists $z \in K$ such that

$$(z) = \frac{(z)_0}{\mathfrak{o}\mathfrak{o}'}.$$

Since the divisor of poles of z has degree 2, we have $[K : k(z)] = 2$ and there is an automorphism of K of order 2 fixing $k(z)$; we shall denote this automorphism by $\sigma_{\mathfrak{o},\mathfrak{o}'}$ and call it the *symmetry* associated with $\mathfrak{o}$ and $\mathfrak{o}'$.

Let us put $\sigma = \sigma_{\mathfrak{o},\mathfrak{o}'}$; since σ fixes z, $\mathfrak{o}^\sigma$ must be $\mathfrak{o}$ or $\mathfrak{o}'$. Now for any divisor $\mathfrak{a}$, $\mathfrak{a}\mathfrak{a}^\sigma$ is a divisor of $k(z)$. In particular, $\mathfrak{o}\mathfrak{o}^\sigma$ is a divisor of $k(z)$, but $\mathfrak{o}^2$ is not, unless $\mathfrak{o}' = \mathfrak{o}$. Hence if $\mathfrak{o}' \neq \mathfrak{o}$, then $\mathfrak{o}^\sigma = \mathfrak{o}'$, $\mathfrak{o}'^\sigma = \mathfrak{o}$.

If $\mathfrak{p}$ is any place of K, then $\dfrac{\mathfrak{p}}{\mathfrak{o}} \cdot \dfrac{\mathfrak{p}^\sigma}{\mathfrak{o}^\sigma}$ is of degree 0 in $k(z)$ and hence is principal:

$$\frac{\mathfrak{p}}{\mathfrak{o}} \cdot \frac{\mathfrak{p}^\sigma}{\mathfrak{o}^\sigma} \sim \mathfrak{e}, \quad \text{hence} \quad \frac{\mathfrak{p}}{\mathfrak{o}} \cdot \frac{\mathfrak{p}^\sigma}{\mathfrak{o}} \sim \frac{\mathfrak{o}'}{\mathfrak{o}}.$$

This may be written as $\mathfrak{p}+\mathfrak{p}^{\sigma}=\mathfrak{o}'$ or

$$\mathfrak{p}^{\sigma}=\mathfrak{o}'-\mathfrak{p},\quad \text{where } \sigma=\sigma_{\mathfrak{o},\mathfrak{o}'} \text{ and } \mathfrak{o}'\neq\mathfrak{o}. \tag{5}$$

Here $\mathfrak{o}$ is understood to be the neutral element for addition. The result still holds for $\mathfrak{o}'=\mathfrak{o}$; in that case it simplifies to

$$\mathfrak{p}^{\sigma}=-\mathfrak{p},\quad \text{where } \sigma=\sigma_{\mathfrak{o},\mathfrak{o}}. \tag{6}$$

We shall define the *translation* associated with $\mathfrak{o}$, $\mathfrak{o}'$ by

$$\tau_{\mathfrak{o},\mathfrak{o}'}=\sigma_{\mathfrak{o},\mathfrak{o}}\sigma_{\mathfrak{o},\mathfrak{o}'}. \tag{7}$$

Writing $\tau=\tau_{\mathfrak{o},\mathfrak{o}'}$, for short, we have

$$\mathfrak{p}^{\tau}=\mathfrak{o}'-(-\mathfrak{p})=\mathfrak{p}+\mathfrak{o}'.$$

If $\tau=\tau_{\mathfrak{o},\mathfrak{o}'}$, $\tau'=\tau_{\mathfrak{o},\mathfrak{o}''}$, then

$$\mathfrak{p}^{\tau\tau'}\sim\left(\frac{\mathfrak{p}\mathfrak{o}'}{\mathfrak{o}}\right)^{\tau'}\sim\frac{\mathfrak{p}\mathfrak{o}''}{\mathfrak{o}}\cdot\frac{\mathfrak{o}'\mathfrak{o}''}{\mathfrak{o}}\cdot\frac{\mathfrak{o}}{\mathfrak{o}\mathfrak{o}''}\sim\frac{\mathfrak{p}\mathfrak{o}''\mathfrak{o}'}{\mathfrak{o}^2};$$

thus

$$\mathfrak{p}^{\tau\tau'}=\mathfrak{p}+\mathfrak{o}'+\mathfrak{o}''.$$

It follows that

$$\tau_{\mathfrak{o},\mathfrak{o}'}\tau_{\mathfrak{o},\mathfrak{o}''}=\tau_{\mathfrak{o},\mathfrak{o}'+\mathfrak{o}''}. \tag{8}$$

For we have seen that both sides have the same effect on all places. What this shows is that the translations (with a given place $\mathfrak{o}$ as neutral) form a group isomorphic to D_0, the group of divisors of degree zero, via the isomorphism

$$\frac{\mathfrak{o}'}{\mathfrak{o}}\mapsto\tau_{\mathfrak{o},\mathfrak{o}'}.$$

We can use this mapping to give a description of Aut (K/k):

Theorem 7.4

Let G be the group of all automorphisms of an elliptic function field K/k, denote by S_0 the stabilizer of a place $\mathfrak{o}$ on K and by T the group of all translations with $\mathfrak{o}$ as neutral. Then T is normal in G and G is the semidirect product of S_o and T:

$$G=S_0T,\qquad S_0\cap T=1.$$

Proof

Let $\tau\in T$, say $\tau=\tau_{\mathfrak{o},\mathfrak{o}'}$ and $\sigma\in G$. Then

$$\left(\frac{\mathfrak{p}}{\mathfrak{o}}\right)^{\sigma^{-1}\tau\sigma}=\left(\frac{\mathfrak{p}^{\sigma^{-1}}}{\mathfrak{o}^{\sigma^{-1}}}\right)^{\tau\sigma}=\left(\frac{\mathfrak{p}^{\sigma^{-1}}\mathfrak{o}'}{\mathfrak{o}^{\sigma^{-1}}\mathfrak{o}}\right)^{\sigma}=\frac{\mathfrak{p}}{\mathfrak{o}}\cdot\frac{\mathfrak{o}'^{\sigma}}{\mathfrak{o}^{\sigma}}.$$

Hence

$$\sigma^{-1}\tau_{\mathfrak{o},\mathfrak{o}'}\sigma = \tau_{\mathfrak{o}^{\sigma},\mathfrak{o}'^{\sigma}} = \tau_{\mathfrak{o},\mathfrak{o}'^{\sigma}-\mathfrak{o}^{\sigma}}. \tag{9}$$

This equation describes the action of G on T and it shows that T is normal in G. If $\tau = \tau_{\mathfrak{o},\mathfrak{o}'}$ fixes $\mathfrak{o}$, then $\mathfrak{o} = \mathfrak{o}^{\tau} = \mathfrak{o} + \mathfrak{o}' = \mathfrak{o}'$, hence $\tau = \tau_{\mathfrak{o},\mathfrak{o}} = 1$; therefore $S_0 \cap T = 1$. Finally, given $\sigma \in G$, if $\mathfrak{o}^{\sigma} = \mathfrak{o}'$, then $\sigma\tau_{\mathfrak{o},\mathfrak{o}'}^{-1} \in S_0$, hence $G = S_0 T$. ■

Later (in Corollary 7.10) we shall see that the stabilizer S_0 is a finite group (it is cyclic of order at most 6).

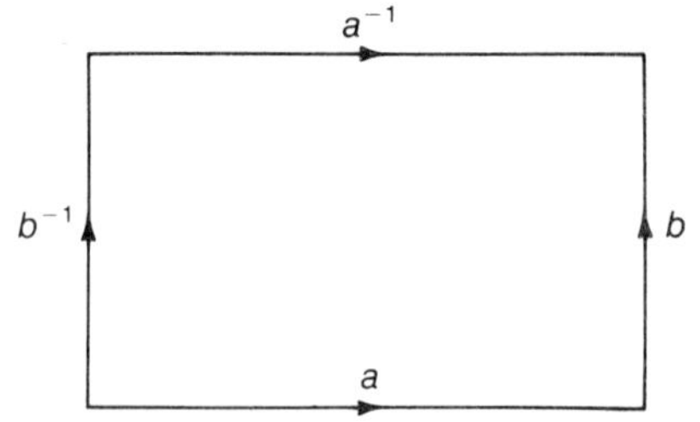

Let us consider more particularly the case $k = \mathbf{C}$. We recall from section 4.4 that the canonical form of a Riemann surface of genus 1 is a torus, which may be represented by a rectangle with opposite sides identified. By the Riemann–Roch theorem, $N(W) = g = 1$; this means that there is just one differential ϕ with integral divisor (viz. $\mathfrak{e}$). To find it, fix a place $\mathfrak{o}$ and write z for a uniformizer at $\mathfrak{o}$. As we saw earlier, there is a function $w \in K$ with poles in $\mathfrak{o}^2$, and as our function with poles in $\mathfrak{o}^3$ but not in $\mathfrak{o}^2$ we may take dw/dz. By the reduction given earlier we have

$$\left(\frac{dw}{dz}\right)^2 = 4w^3 - g_2 w - g_3.$$

It follows that

$$z = \int \frac{dw}{\sqrt{(4w^3 - g_2 w - g_3)}}, \tag{10}$$

and here the integrand may be taken as our differential ϕ. For this is determined up to a scalar (because W is one-dimensional) and

$$dz = (4w^3 - g_2 w - g_3)^{-1/2} dw$$

remains finite at the zeros of $4w^3 - g_2 w - g_3$, as an elementary integration shows, while at $\mathfrak{o}$ the pole of this function just cancels that of dw. The integral (10) is also called an *elliptic integral*. An integral of this kind can be used to express the arc length of an ellipse; this accounts for the name.

At any place $\mathfrak{p}$ the function z takes the value

$$z(\mathfrak{p}) = \int_{\mathfrak{o}}^{\mathfrak{p}} \phi.$$

If t is a uniformizer at $\mathfrak{p}$, then $dz/dt \neq 0, \infty$, hence the function $z = z(\mathfrak{p})$ maps the Riemann surface conformally on a (curvilinear) quadrangle F in the z-plane. To the closed paths a, b there correspond constants

$$\omega_a = \int_a \phi, \qquad \omega_b = \int_b \phi,$$

called the *periods*. We remark that z, $z + \omega_a$, $z + \omega_b$ all correspond to the same value of w; in other words, w as a function of z is doubly periodic. The subgroup of the additive group of $\mathbf{C}$ generated by ω_a, ω_b is denoted by Ω and is called the *period lattice*. The quadrangle F is a transversal for Ω in $\mathbf{C}$, called a *fundamental domain*. For a suitable choice of coordinate F may be taken as a parallelogram in the form

$$\{\lambda\omega_a + \mu\omega_b \mid 0 \leqslant \lambda, \mu < 1\}.$$

Translating F by $r\omega_a + s\omega_b$ $(r, s \in \mathbf{Z})$ we obtain a paving of the z-plane and in this way we get a covering of the Riemann surface by the plane. We assert that this fills the z-plane just once. No point is covered more than once, for if $z(\mathfrak{p}) = z(\mathfrak{q})$, then $\int_{\mathfrak{p}}^{\mathfrak{q}}\phi = 0$. Let $\mathfrak{r}$ describe a path $U : \mathfrak{p} \rightarrow \mathfrak{q}$; then $\int_{\mathfrak{p}}^{\mathfrak{r}}\phi$ describes a closed path C from 0 to 0 in the z-plane. This can be deformed to 0 and it deforms U to a path $U' : \mathfrak{p} \rightarrow \mathfrak{q}$ which makes $\int_{\mathfrak{p}}^{\mathfrak{r}}\phi = 0$ for all places $\mathfrak{r}$. But this would mean that $z(\mathfrak{r}) = 0$ for all $\mathfrak{r}$ on U', which contradicts the fact that $\phi \neq 0$. The fact that $\phi \neq 0$ also shows that F has a non-empty interior, i.e. ω_a/ω_b is not real, so it is a neighbourhood of some point, and by translation the whole plane gets covered. It is clear that the quotient $\mathbf{C}/\Omega$ is a torus and the mapping from the z-plane to the Riemann surface gives a homeomorphism $\mathbf{C}/\Omega \cong D_0$. We shall show this to be a group isomorphism; this is a special case of the Abel–Jacobi theorem. As a first step we show that differentials are translation-invariant:

Proposition 7.5
Let K/k be an elliptic function field and ϕ a differential of the first kind. Then for any translation τ, $\phi^{\tau} = \phi$.

Proof
The space $D(\mathfrak{e})$ of differentials of the first kind is one-dimensional, so if we take any non-zero differential ϕ in $D(\mathfrak{e})$ and apply an automorphism σ, we find

$$\phi^{\sigma} = \mu_{\sigma}\phi, \quad \text{where } \mu_{\sigma} \in k^{\times}, \tag{11}$$

because ϕ^σ is again in $D(\mathfrak{e})$. Now a symmetry σ satisfies $\sigma^2 = 1$, and in this case (11) shows that $\mu_\sigma^2 = 1$, hence $\mu_\sigma = \pm 1$. It follows that for a translation τ, $\mu_\tau = \pm 1$, so the result will follow if we express any translation as the square of another: if $\tau = \alpha^2$, then $\mu_\tau = \mu_\alpha^2 = 1$. Thus, given places $\mathfrak{o}$, $\mathfrak{p}$, we have to find $\mathfrak{o}'$ such that

$$\frac{\mathfrak{p}}{\mathfrak{o}} \sim \frac{\mathfrak{o}'^2}{\mathfrak{o}^2}, \quad \text{i.e. } \mathfrak{p}\mathfrak{o} \sim \mathfrak{o}'^2.$$

Remembering that $W = E$, we have by the Riemann–Roch formula

$$l((\mathfrak{o}\mathfrak{p})^{-1}) + d((\mathfrak{o}\mathfrak{p})^{-1}) + 1 - 1 = l(\mathfrak{o}\mathfrak{p}) = 0.$$

It follows that $l((\mathfrak{o}\mathfrak{p})^{-1}) = 2$. If we take 1, x linearly independent in $L((\mathfrak{o}\mathfrak{p})^{-1})$, then dx has degree 0 and poles at $\mathfrak{o}$, $\mathfrak{p}$. Let $\mathfrak{o}'$ be a zero of dx; then x is finite at $\mathfrak{o}'$ and its value is α, say. Since dx has a zero at $\mathfrak{o}'$, $x - \alpha$ has a double zero at $\mathfrak{o}'$, so $(x - \alpha) = \dfrac{\mathfrak{o}'^2}{\mathfrak{o}\mathfrak{p}}$; hence $\mathfrak{o}\mathfrak{p}/\mathfrak{o}'^2 \in E$, and this is what we wished to show. ■

As an immediate consequence we have the *addition theorem for integrals of the first kind*:

Theorem 7.6
For any integral of the first kind $\int \phi$ *in an elliptic function field,*

$$\int_{\mathfrak{o}}^{\mathfrak{p}} \phi + \int_{\mathfrak{o}}^{\mathfrak{q}} \phi \equiv \int_{\mathfrak{o}}^{\mathfrak{p}+\mathfrak{q}} \phi \pmod{\Omega},$$

where Ω *is the period lattice and the addition of places is defined as in (2).*

Proof
If $\tau = \tau_{\mathfrak{o},\mathfrak{p}}$, then $\mathfrak{o}^\tau = \mathfrak{p}$, $\mathfrak{q}^\tau = \mathfrak{p} + \mathfrak{q}$; since differentials are translation-invariant, we have

$$\int_{\mathfrak{o}}^{\mathfrak{q}} \phi = \int_{\mathfrak{o}}^{\mathfrak{q}} \phi^{\tau^{-1}} = \int_{\mathfrak{o}^\tau}^{\mathfrak{q}^\tau} \phi \equiv \int_{\mathfrak{p}}^{\mathfrak{p}+\mathfrak{q}} \phi \pmod{\Omega}.$$

Therefore

$$\int_{\mathfrak{o}}^{\mathfrak{p}} \phi + \int_{\mathfrak{o}}^{\mathfrak{q}} \phi \equiv \int_{\mathfrak{o}}^{\mathfrak{p}} \phi + \int_{\mathfrak{p}}^{\mathfrak{p}+\mathfrak{q}} \phi = \int_{\mathfrak{o}}^{\mathfrak{p}+\mathfrak{q}} \phi \pmod{\Omega}. \blacksquare$$

Let us note for reference that a differential or the corresponding integral is said to be of the *second kind* if it has poles but no residues, and of the *third kind* if it has poles with non-zero residues. We remark that for an elliptic function field a differential ω is of the second kind if and only if for each place $\mathfrak{p}$ there is a function u such that $v_{\mathfrak{p}}(\omega - du) \geqslant 0$, but for some $\mathfrak{p}$, $v_{\mathfrak{p}}(\omega) < 0$.

We saw in Theorem 7.1 that the additive group of places is isomorphic to

D_0, so by Theorem 7.6 the map $\mathfrak{p} \mapsto \int_{\mathfrak{o}}^{\mathfrak{p}} \phi$ provides a homomorphism from D_0 to $\boldsymbol{C}/\Omega$. The image is a subgroup including a neighbourhood of 0, and hence is the whole group, because $\boldsymbol{C}/\Omega$ is connected. To show that it is an isomorphism it remains to prove injectivity. This follows from:

Theorem 7.7 (*Abel's theorem*)
Let K/k be an elliptic function field. A divisor of degree zero,

$$\mathfrak{a} = \frac{\mathfrak{p}_1 \ldots \mathfrak{p}_n}{\mathfrak{q}_1 \ldots \mathfrak{q}_n}$$

is principal if and only if

$$\sum_1^n \int_{\mathfrak{q}_i}^{\mathfrak{p}_i} \phi \equiv 0 \pmod{\Omega},$$

where ϕ is a non-zero differential of the first kind and Ω is the period lattice.

Proof
We have to show that $\mathfrak{p}_1 \ldots \mathfrak{p}_n \mathfrak{o}^{-n} \sim \mathfrak{q}_1 \ldots \mathfrak{q}_n \mathfrak{o}^{-n}$ if and only if

$$\sum_1^n \int_{\mathfrak{o}}^{\mathfrak{p}_i} \phi \equiv \sum_1^n \int_{\mathfrak{o}}^{\mathfrak{q}_i} \phi \pmod{\Omega}.$$

Write $\mathfrak{p} = \Sigma \mathfrak{p}_i$, $\mathfrak{q} = \Sigma \mathfrak{q}_i$; then by Theorem 7.6 we must show that

$$\mathfrak{p}\mathfrak{o}^{-1} \sim \mathfrak{q}\mathfrak{o}^{-1} \Leftrightarrow \int_{\mathfrak{o}}^{\mathfrak{p}} \phi \equiv \int_{\mathfrak{o}}^{\mathfrak{q}} \phi \pmod{\Omega},$$

and clearly the right-hand side holds if and only if

$$\int_{\mathfrak{p}}^{\mathfrak{q}} \phi \equiv 0 \pmod{\Omega}. \tag{12}$$

But we saw that $\mathfrak{p}\mathfrak{o}^{-1} \sim \mathfrak{q}\mathfrak{o}^{-1} \Leftrightarrow \mathfrak{p} = \mathfrak{q}$. If we choose paths from $\mathfrak{o}$ to $\mathfrak{p}$ and to $\mathfrak{q}$ inside a fundamental domain, then the congruence (12) becomes an equation, and this holds precisely for $\mathfrak{p} = \mathfrak{q}$.■

We shall digress briefly at this point to show how Picard's theorem can be proved very simply from the results obtained here. This theorem (and its proof) was given by Picard in 1880 (at the age of 24); later other more complicated (but more elementary) proofs were found (e.g. using Bloch's theorem).

The well-known theorem of Casorati–Weierstrass tells us that an analytic function near an essential singularity comes arbitrarily close to every value. However, it need not assume every value; thus e^z never assumes the value 0, nor ∞, since the function is undefined for $z = \infty$. But a function which omits three values must be a constant. This is Picard's theorem, and (in outline) his proof runs as follows.

Let us consider the integral (10). Taking the zeros of the cubic in the denominator to be 0, 1, λ, we have a function

$$z = \int \frac{dw}{\sqrt{[w(w-1)(w-\lambda)]}}.$$

It can be shown that the two periods ω_1, ω_2 of this integral are analytic functions of λ (this is a verification which we shall omit, cf. Picard (1893) Vol. II, p. 227). These periods are defined for any $\lambda \neq 0, 1, \infty$; hence their ratio $j(\lambda) = \omega_2/\omega_1$ is also an analytic function, called the *elliptic modular function*. As we have seen, $j(\lambda)$ is never real, hence as λ varies, $j(\lambda)$ cannot cross the real axis, so the values of $j(\lambda)$ lie in a half-plane, say the upper half-plane. There is an analytic function mapping the upper half-plane to the interior of the unit circle, e.g.

$$f(z) = \frac{z-i}{-iz+1}.$$

Suppose now that $g(z)$ is an analytic function which omits three values, say 0, 1, ∞, as z varies over the Riemann sphere. Then $j(g(z))$ takes values only in the upper half-plane, hence $f(j(g(z)))$ takes values in the unit circle. Thus it is bounded and so, by Liouville's theorem, a constant. It follows that g is a constant and this completes the proof.

We conclude this section with some remarks on the classification of elliptic function fields. We recall that any elliptic function field can be defined by an equation in the Weierstrass normal form

$$y^2 = 4x^3 - g_2x - g_3, \quad g_2, g_3 \in k. \tag{13}$$

Its discriminant $g_2^3 - 27g_3^2$ is not zero because our field is not rational. Therefore the function

$$h(K) = 12^3 \frac{g_2^3}{g_2^3 - 27g_3^2} \tag{14}$$

is always finite. Let us make a substitution

$$x \mapsto c^2x, \qquad y \mapsto c^3y, \quad \text{where } c \in k^\times. \tag{15}$$

This gives an equation of the same form as (13); if $x' = c^2x$, $y' = c^3y$, then (13) becomes

$$y'^2 = c^6y^2 = 4x'^3 - c^4g_2x' - c^6g_3.$$

Hence the constants in (13) undergo the transformation

$$g_2 \mapsto c^4g_2, \qquad g_3 \mapsto c^6g_3. \tag{16}$$

In particular, the transformation (16) leaves (14) unchanged. In fact, these are the only such transformations:

Proposition 7.8
Let K/k be an elliptic function field with Weierstrass normal form (13) for its generators x, y. This form is determined by K/k up to substitutions (15) and they leave invariant the element $h(K)$ given by (14).

Proof
Let

$$y'^2 = 4x'^3 - g'_2 x' - g'_3 \tag{17}$$

be another generation of K/k in Weierstrass normal form. Then $(x') = \mathfrak{x}'/\mathfrak{o}'^2$, $(x) = \mathfrak{x}/\mathfrak{o}^2$ say. The symmetry $\sigma_{\mathfrak{o},\mathfrak{o}'}$ sends x' to $\alpha x + \beta$ ($\alpha \neq 0$), so we may assume

$$x' = \alpha x + \beta, \quad \text{where } \alpha \neq 0.$$

Then y' has poles within $\mathfrak{o}^3$ and so is a linear combination of 1, x, y:

$$y' = \gamma x + \delta y + \varepsilon, \quad \text{where } \delta \neq 0.$$

Inserting these values in (17), we find

$$\gamma^2 x^2 + \delta^2 y^2 + \varepsilon^2 + 2\gamma\varepsilon x + 2\delta\varepsilon y + 2\gamma\delta xy =$$

$$4(\alpha^3 x^3 + 3\alpha^2\beta x^2 + 3\alpha\beta^2 x + \beta^3) - g_2'(\alpha x + \beta) - g_3'.$$

This differs from (13) only by a constant factor, so on comparing coefficients of xy, y, x^2 we find $\gamma = 0$, $\varepsilon = 0$, $\beta = 0$. Hence $x' = \alpha x$, $y' = \delta y$ and $\delta^2 = \alpha^3$. Writing $c = \delta/\alpha$, we find that $x' = c^2 x$, $y' = c^3 y$, as claimed; now the rest is clear.■

The next question is whether an elliptic function field exists for every value of $h(K)$. This is answered affirmatively as follows. Let $c \in k^\times$; to construct K with $h(K) = c$ we distinguish three cases:

(i) $c = 1728$ ($= 12^3$). Then $g_3 = 0$, $y^2 = 4(x^3 - x)$. This is known as the *lemniscate case*, since it occurs in calculating the arc of the lemniscate

$$(x^2 + y^2)^2 = a^2(x^2 - y^2).$$

(ii) $c = 0$. Then $g_2 = 0$, $y^2 = 4(x^3 - 1)$. This is called the *anharmonic case*.

(iii) $c \neq 0$, 1728. Write $\gamma = c - 12^3$ and consider the equation

$$y^2 = 4x^3 - 3c\gamma x - c\gamma^2.$$

Here $g_2 = 3c\gamma$, $g_3 = c\gamma^2$, hence

$$g_2^3 - 27g_3^2 = 27c^3\gamma^3 - 27c^2\gamma^4 = 27c^2\gamma^3(c - \gamma) = 3^3 \,.\, 12^3 c^2\gamma^3$$

and

$$h(K) = 12^3 \,.\, 27c^3\gamma^3/3^3 12^3 c^2\gamma^3 = c.$$

Thus we have for each value of c at least one field with invariant $h(K) = c$. In fact, for any value of c there is precisely one such field up to isomorphism:

Theorem 7.9

Let K, K' be elliptic function fields over a field k, (algebraically closed of characteristic 0), such that $h(K) = h(K')$. Then $K \cong K'$.

Proof

Suppose first that $h(K) = h(K') \neq 0$. By a substitution (15) with suitable c we can arrange that $g_2' = g_2$; comparing values of h we find that $g_3'^2 = g_3^2$ and so $g_3' = \pm g_3$. If $g_3' = -g_3$ we again apply (15) with $c = -1$. By (16) this leaves g_2 unchanged and changes the sign of g_3, so the defining equations for K, K' now have the same form, whence $K \cong K'$.

When $h = 0$, then $g_2 = g_2' = 0$ and we can apply a transformation (15) so as to obtain $g_3' = g_3$.■

The arguments leading to this result can also be used to determine the stabilizer of a place. Given any place $\mathfrak{p}$ of an elliptic function field K/k, an automorphism fixing $\mathfrak{p}$ must have the form (15) and moreover, c must satisfy $g_2 = c^4 g_2$, $g_3 = c^6 g_3$. Again there are three cases:

(i) $h = 12^3$. Then $g_3 = 0$, $g_2 \neq 0$, so $c^4 = 1$ and we have a cyclic group of order 4.

(ii) $h = 0$. Then $g_2 = 0$, $g_3 \neq 0$, so $c^6 = 1$ and we have a cyclic group of order 6.

(iii) $h \neq 12^3, 0$. Then $c^4 = c^6 = 1$ hence $c^2 = 1$ and we have a cyclic group of order 2. Thus we have proved:

Corollary 7.10

Let K/k be an elliptic function field. Then the automorphism group is transitive on the set of places and the stabilizer of a place is a cyclic group of order m, where $m = 4$ if $h = 12^3$, $m = 6$ if $h = 0$ and $m = 2$ otherwise.■

For function fields of genus greater than 1 the situation is entirely different; here the whole automorphism group is finite:

Theorem 7.11

Any function field of genus $g > 1$ has a finite automorphism group.

Proof

We recall from Theorem 6.11 that a function field of genus $g > 1$ has finitely many Weierstrass points, at least $2(g+1)$ in number. Since these places are invariants of K, it follows that any automorphism σ permutes all the Weierstrass points. Suppose that σ fixes every Weierstrass point but moves a place $\mathfrak{p}$ to $\mathfrak{p}' \neq \mathfrak{p}$. Then $\mathfrak{p}$, $\mathfrak{p}'$ cannot correspond to Weierstrass points, so the gaps occur in a block at the beginning and there is a function $x \in K$ with $\mathfrak{p}^{g+1}$ as divisor of poles. Hence $x - x^{\sigma}$ has as its divisor of poles $(\mathfrak{p}\mathfrak{p}')^{g+1}$, of degree $2(g+1)$, so its divisor of zeros is also of degree $2(g+1)$. Any Weierstrass point is fixed by σ, so x and x^{σ} assume the same value there, and hence the difference $x - x^{\sigma}$ has a zero at each Weierstrass point. But for a function field which is not hyperelliptic the number of Weierstrass points is $> 2(g+1)$, so $x - x^{\sigma}$ must vanish identically, i.e. $x^{\sigma} = x$ and $\sigma = 1$. This shows that the automorphism group of K/k is a faithful permutation group on the finite set of Weierstrass points, unless K is hyperelliptic. If K is hyperelliptic but not elliptic, then there is a function z whose divisor of poles has degree 2, hence $z - z^{\sigma}$ has a divisor of poles of degree at most 4. Now $g \geqslant 2$, hence $2g + 2 \geqslant 6$ and we conclude as before that $z^{\sigma} = z$. Since $[K : k(z)] = 2$, there are at most two automorphisms leaving $k(z)$ fixed, so the automorphism group of K is again finite, as claimed.■

A closer examination of this proof will give an explicit bound for the order, but this is far from the best possible. Hurwitz has shown that a field of genus $g > 1$ has at most $84(g-1)$ automorphisms and this bound is attained for certain fields of genus 3.

Exercises

1. Carry out the details in the proof of Theorem 7.2 for the case when two or more of $\mathfrak{o}$, $\mathfrak{p}_1$, $\mathfrak{p}_2$, $\mathfrak{p}_3$ coincide.
2. Show how an integral $\int f^{-1/2} dw$, where f is a quartic polynomial in w, can be reduced to the form (10), by transforming one of the zeros to ∞. (*Hint*: Use the remarks after Theorem 6.9.)
3. Verify that in an elliptic function field a differential ω is of the first or second kind if and only if at any point $\mathfrak{p}$ there is a function u such that $v_{\mathfrak{p}}(\omega - du) \geqslant 0$. Can this ever hold for function fields that are not rational nor elliptic?
4. Show that the group of automorphisms of a function field K/k of genus 0 (over an algebraically closed ground field) is isomorphic to $\mathbf{PGL}_2(k)$, the projective linear group, and the stabilizer of a place is isomorphic to the affine linear group $x \mapsto ax + b$. What is the stabilizer of a pair of distinct places?

4.8 ABELIAN INTEGRALS AND THE ABEL–JACOBI THEOREM

We shall now take our constant field to be $\mathbf{C}$. This will allow us to use topological ideas, although much of the argument could be carried out for more general cases by making appropriate definitions.

Let $K/\mathbf{C}$ be a function field of genus g. We recall from section 4.4 that the corresponding Riemann surface R is represented by a polygon with $2g$ sides $a_1, \ldots, a_g, b_1, \ldots, b_g$ whose vertices are all identified. For simplicity we shall write $\tilde{a}_r = a_{g+r} = b_r, \tilde{a}_{g+r} = -a_r (r = 1, \ldots, g)$. For $g = 1$ we have a parallelogram, and as we have seen in section 4.7, the universal covering surface is the plane. For $g = 0$ we have of course a rational function field and the universal covering surface is then the whole Riemann sphere. It can further be shown that for $g > 1$ the universal covering surface is the unit circle (Koebe's uniformization theorem, 1908). These cases are also known as the elliptic ($g = 0$), parabolic ($g = 1$) and hyperbolic case ($g > 1$).

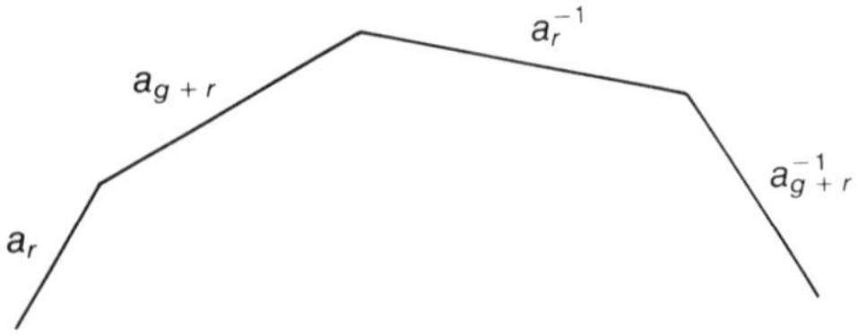

The space $D(\mathfrak{e})$ of differentials of the first kind (i.e. without poles) is g-dimensional (Corollary 6.3); we shall take a fixed basis $\phi_1, \ldots, \phi_g$ and put

$$\omega_{r\nu} = \int_{a_r} \phi_\nu. \tag{1}$$

The numbers $\omega_{1\nu}, \ldots, \omega_{2g\nu}$ are called the *periods* of ϕ_ν and $\Omega = (\omega_{r\nu})$ is a $2g \times g$ matrix called the *period matrix*. We shall further put $\tilde{\omega}_{r\nu} = \omega_{g+r\nu}$, $\tilde{\omega}_{g+r\nu} = -\omega_{r\nu}$.

A differential may be visualized as a path element; thus $\int \phi$ is defined along paths. Likewise, products of differentials represent elements of area and $\int \phi \wedge \psi$ is defined over an area. Here orientation is taken into account, thus $\psi \wedge \phi = -\phi \wedge \psi$ and of course $\phi \wedge \phi = 0$. We shall not formalize these notions but trust that the reader will remember enough vector calculus to follow the argument. In particular, we recall Stokes' theorem: let z be any function in the plane, θ a differential and A a plane region with suitably well-behaved boundary ∂A. Then

$$\int_A dz \wedge \theta = \int_{\partial A} z\theta.$$

We also recall Cauchy's residue theorem: the integral of an analytic differential θ around a closed curve is $2\pi i$ times the sum of the residues of θ at its poles inside the curve.

As a first step we show how to express an integral in terms of the periods.

Lemma 8.1

Let K be a function field and $\phi_1, \ldots, \phi_g$ a basis for the space of differentials of the first kind, with period matrix given by (1). Further, let θ be a differential *(possibly with poles, but none on the boundary of the fundamental polygon), and put*

$$\lambda_r = \int_{a_r} \theta$$

for its periods. Then

$$\int \phi_\nu \wedge \theta = 2\pi i \,.\, \sum \mathrm{res}_\mathfrak{p}\left(\theta \int_\mathfrak{o}^\mathfrak{p} \phi_\nu\right) = \sum \lambda_r \tilde{\omega}_{r\nu}, \tag{2}$$

where the integral on the left is taken over the fundamental polygon.

Proof

Write $f_\nu(\mathfrak{p}) = \int_\mathfrak{o}^\mathfrak{p} \phi_\nu$, taken along a path inside the fundamental polygon. By Stokes' theorem,

$$\int \phi_\nu \wedge \theta = \int f_\nu \theta = 2\pi i \,.\, \sum \mathrm{res}_\mathfrak{p}(f_\nu(\mathfrak{p})\theta).$$

To evaluate this sum, we integrate along the boundary:

along a_r: $-\int_{a_r} f_\nu\theta + \int_{a_r} (f_\nu + \omega_{g+r\,\nu})\theta = \lambda_r\omega_{g+r\,\nu} = \lambda_r\tilde{\omega}_{r\,\nu}$;

along a_{g+r}: $-\int_{a_{g+r}} f_\nu\theta + \int_{a_{g+r}} (f_\nu - \omega_{r\nu})\theta = -\lambda_{g+r}\omega_{r\,\nu} = \lambda_{g+r}\tilde{\omega}_{g+r\,\nu}$.

Hence the total is $\Sigma\,\lambda_r\tilde{\omega}_{r\nu}$ and we obtain (2).■

Integrals over areas satisfy an important reality condition. We shall use the bar to indicate complex conjugates, as usual.

Lemma 8.2

Let K be any function field. For any $\phi, \psi \in D(\mathfrak{e})$,

$$\int \phi \wedge \psi = 0, \tag{3}$$

$$-i\int \bar{\phi} \wedge \phi \geqslant 0, \quad \textit{with equality if and only if } \phi = 0. \tag{4}$$

Proof

Equation (3) follows by Lemma 8.1, because ϕ has no poles. To prove (4) we take a uniformizer $z = x + iy$ and write $\phi = w\,dz$, where $w = u + iv$; we have $\bar{\phi} \wedge \phi = |w|^2(dx - i\,dy) \wedge (dx + i\,dy) = 2i\,|w|^2 dx \wedge dy$. Now $dx \wedge dy = dA$ is an element of area, which in the customary orientation is positive, so $-i\bar{\phi} \wedge \phi = 2\,|w|^2 dA \geqslant 0$, and the inequality in (4) follows. If we had equality, then $w = 0$ everywhere, by continuity, so $\phi = 0$ as asserted.■

We remark that in terms of the period matrix Ω we have $(\tilde{\omega}_{r\nu}) = J\Omega$, where

$$J = \begin{pmatrix} 0 & I \\ -I & 0 \end{pmatrix}.$$

Given any matrix A, let us write A^T for the transpose and $A^H = \bar{A}^T$ for the Hermitian conjugate (conjugate transpose). With this notation the result of Lemma 8.2 can be restated in terms of matrices, as follows:

Theorem 8.3
Let K be a function field of genus g. Denote the period matrix by Ω and write

$$J = \begin{pmatrix} 0 & I \\ -I & 0 \end{pmatrix}.$$

Then

$$\Omega^{\mathrm{T}} J \Omega = 0 \quad \text{(Riemann's equality)}, \tag{5}$$

$$-i\Omega^{\mathrm{H}} J\Omega \textit{ is Hermitian positive definite} \text{ (Riemann's inequality)}. \tag{6}$$

Proof
In terms of our basis for $D(\mathfrak{e})$ we can write

$$\phi = \Sigma\, \alpha_\nu \phi_\nu,$$

$$\psi = \sum \beta_\nu \phi_\nu \quad (\alpha_\nu, \beta_\nu \in \mathbf{C}),$$

for any differentials in $D(\mathfrak{e})$. The periods of ψ are then $\mu_r = \Sigma\, \omega_{r\nu}\beta_\nu$ and by Lemmas 8.1 and 8.2 we have

$$0 = \int \phi \wedge \psi = \sum \alpha_\nu \int \phi_\nu \wedge \psi = \sum \alpha_\nu \,.\, \sum \mu_r \tilde{\omega}_{r\,\nu} = \sum \alpha_\nu \tilde{\omega}_{r\,\nu} \omega_{r\rho} \beta_\rho.$$

Thus, writing $\alpha = (\alpha_1, \ldots, \alpha_g)^T$, $\beta = (\beta_1, \ldots, \beta_g)^T$, we find that

$$\alpha^{\mathrm{T}} \Omega^{\mathrm{T}} J \Omega \beta = -(J\Omega\alpha)^T \Omega\beta = 0.$$

Since α, β were arbitrary, (5) follows.

To prove (6) we note that the matrix $A = -i\Omega^{\mathrm{H}} J\Omega$ is Hermitian, since $A^H = i\Omega^{\mathrm{H}} J^H \Omega = -i\Omega^{\mathrm{H}} J\Omega = A$. Now if $\phi = \Sigma\, \alpha_\nu \phi_\nu$, then

$$-i\alpha^{\mathrm{H}} \Omega^{\mathrm{H}} J \Omega \alpha = -i\int \bar{\phi} \wedge \phi,$$

and by Lemma 8.2 this is $\geqslant 0$, with equality only when $\alpha = 0$.■

It follows from (6) that Ω has rank g. Any matrix Ω satisfying (5), (6) is called a *Riemann matrix*; more generally, (Ω, X) is a *Riemann matrix pair* if X is a skew-symmetric $2g \times 2g$ integer matrix and (5), (6) hold with X in place of J. Further, two such pairs (Ω, X) and (Ω', X') are said to be *equivalent* if there exists an invertible $2g \times 2g$ matrix A over $\mathbf{Z}$ and an invertible $g \times g$ matrix B over $\mathbf{C}$ such that

$$\Omega' = A^{-1}\Omega B, \;\; X' = A^T X A. \tag{7}$$

Clearly a change of basis in $D(\mathfrak{e})$ is accomplished by an invertible matrix over $\mathbf{C}$ (corresponding to B in (7)), while a transformation by A corresponds to a change in the fundamental polygon. We can use the freedom in the choice of A and B to reduce Ω as follows.

Write

$$\Omega = \begin{pmatrix} \Omega_1 \\ \Omega_2 \end{pmatrix},$$

where Ω_1, Ω_2 are $g \times g$ matrices over $\mathbf{C}$. Since the Riemann matrix Ω has rank g, we may assume Ω_2 non-singular, and by a transformation (7) with $B = \Omega_2^{-1}$ we reduce Ω to the form

$$\Omega = \begin{pmatrix} \Omega_0 \\ I \end{pmatrix}. \tag{8}$$

We note that X has been unaffected by this reduction. If two Riemann matrices in the form (8) are equivalent, we have an equation

$$\begin{pmatrix} \Omega_1 \\ I \end{pmatrix} = \begin{pmatrix} A & B \\ C & D \end{pmatrix} \begin{pmatrix} \Omega_0 \\ I \end{pmatrix} M = \begin{pmatrix} (A\Omega_0 + B)M \\ (C\Omega_0 + D)M \end{pmatrix}.$$

It follows that

$$M = (C\Omega_0 + D)^{-1}, \qquad \Omega_1 = (A\Omega_0 + B)(C\Omega_0 + D)^{-1}, \tag{9}$$

and these equations describe the equivalence completely. For example, for $g = 1$ we obtain the unimodular group (i.e. the group of integer matrices with determinant 1). If we insist that $X = J$, we are restricted to the subgroup of the unimodular group leaving J fixed, i.e. the symplectic group (cf. A.1, section 8.6).

Let us put $\Lambda = \Omega^{\mathrm{T}} J$; then $\Lambda\Omega = 0$ and each of Λ, Ω is the exact annihilator of the other over $\mathbf{C}$, by Sylvester's law of nullity, because both have rank g. We have the exact sequence of row vectors

$$0 \to \mathbf{Z}^g \to \mathbf{Z}^{2g} \xrightarrow{\Omega} \mathbf{C}^g \to J \to 0.$$

Here J is a compact abelian group, called the *Jacobian variety*. Similarly we have an exact sequence of column vectors

$$0 \to {}^g\mathbf{Z} \to {}^{2g}\mathbf{Z} \xrightarrow{\Lambda} {}^g\mathbf{C} \to P \to 0.$$

Here P is again a compact abelian group, the *Picard variety*.

Let us again take a fixed basis $\phi_1, \ldots, \phi_g$ of $D(\mathfrak{e})$ and fix a place $\mathfrak{o}$. We have a mapping from the Riemann surface of our function field to $J = \mathbf{C}^g/\mathbf{Z}^{2g}\Omega$, given by

$$\mathfrak{p} \mapsto S(\mathfrak{p}) = \left(\int_{\mathfrak{o}}^{\mathfrak{p}} \phi_1, \ldots, \int_{\mathfrak{o}}^{\mathfrak{p}} \phi_g \right).$$

By linearity this extends to a homomorphism of the divisor group: $D \to J$. If we restrict this mapping to D_0, the group of divisors of degree 0, we obtain a mapping which does not depend on the choice of $\mathfrak{o}$:

$$S:\mathfrak{a} \mapsto S(\mathfrak{a}) = \left(\sum n_{\mathfrak{p}} \int_{\mathfrak{o}}^{\mathfrak{p}} \phi_1, \ldots, \sum n_{\mathfrak{p}} \int_{\mathfrak{o}}^{\mathfrak{p}} \phi_g\right), \quad \text{where } \mathfrak{a} = \prod \mathfrak{p}^{n_{\mathfrak{p}}}. \qquad (10)$$

Because $\mathfrak{a}$ has degree 0, each sum can be rewritten as an integral not involving $\mathfrak{o}$. This mapping is the isomorphism we have been seeking, but to prove this fact we must be able to construct differentials with preassigned residues:

Lemma 8.4
In any function field let $\mathfrak{a} = \Pi\mathfrak{p}^{n_{\mathfrak{p}}}$ be a divisor of degree zero. Then there is a differential θ such that

$$\text{res}_{\mathfrak{p}}\theta = n_{\mathfrak{p}}. \qquad (11)$$

Proof
Suppose first that $\mathfrak{a} = \mathfrak{p}^n\mathfrak{q}^{-n}$. By the Riemann–Roch formula,

$$l((\mathfrak{p}\mathfrak{q})^{-1}) = -d(\mathfrak{p}\mathfrak{q}) + 1 - g + \delta(\mathfrak{p}\mathfrak{q}).$$

Hence $\delta(\mathfrak{p}\mathfrak{q}) = l((\mathfrak{p}\mathfrak{q})^{-1}) + g + 1 > g$, so there exists a differential θ with a pole of order 1 at $\mathfrak{p}$ or $\mathfrak{q}$ and no others. But the sum of the residues is 0, so θ has non-zero residues at $\mathfrak{p}$ and $\mathfrak{q}$, equal and of opposite signs. On multiplying by a suitable constant, we have

$$\text{res}_{\mathfrak{p}}(\theta) = n = -\,\text{res}_{\mathfrak{q}}(\theta),$$

as required.

In general let $\mathfrak{a} = \Pi\mathfrak{p}_\nu^{n_\nu}$, $\Sigma n_\nu = 0$. Pick a place $\mathfrak{o}$ distinct from all the $\mathfrak{p}_\nu$ and for each $\mathfrak{p}_\nu$ construct θ_ν such that

$$\text{res}_{\mathfrak{p}_\nu}(\theta_\nu) = n_\nu = -\,\text{res}_{\mathfrak{o}}(\theta_\nu).$$

Then $\theta = \Sigma\,\theta_\nu$ satisfies the required relation (11).■

Theorem 8.5 (*Abel–Jacobi theorem*)
For any function field $K/\mathbf{C}$ the mapping (10) gives rise to an exact sequence

$$1 \to \mathbf{C}^\times \to K^\times \to D_0 \xrightarrow{S} J \to 0. \qquad (12)$$

Thus $J \cong D_0/F$, where F is the subgroup of principal divisors.

Proof
We begin by showing that ker $S = F$. Take any $z \in K\backslash\mathbf{C}$, say $(z) = \Pi\mathfrak{p}_\nu^{n_\nu}$ and choose the fundamental polygon P so that no pole of z is on its boundary. Write $\zeta = dz/z$; its residue at any point indicates poles (-1) and zeros $(+1)$ with proper multiplicity. Hence for any closed curve not passing through any poles or zeros of z we have $\int \zeta = 2\pi i n$, where n is an integer. Let us write

$$\int_{a_r} \zeta = 2\pi i h_r.$$

Then by Lemma 8.1,

$$\int \phi_\nu \wedge \zeta = 2\pi i \,.\, \sum n_\lambda \int_{\mathfrak{o}}^{\mathfrak{p}_\lambda} \phi_\nu = \sum 2\pi i h_r \tilde{\omega}_{r\,\nu}.$$

Hence

$$S((z)) = \left(\sum n_\lambda \int_{\mathfrak{o}}^{\mathfrak{p}_\lambda} \phi_1, \ldots \right) = \left(\sum h_r \tilde{\omega}_{r\,1}, \ldots \right) \equiv 0 \pmod{\mathbf{Z}^{2g}\Omega}.$$

Conversely, let $\mathfrak{a}$ be a divisor of degree 0, say $\mathfrak{a} = \Pi \mathfrak{p}_\nu^{n_\nu}$, and assume that $S(\mathfrak{a}) = 0$; we have to show that $\mathfrak{a}$ is principal. Since $S(\mathfrak{a}) = 0$, there exist integers m_r such that

$$\sum_\mu n_\mu \int_{\mathfrak{o}}^{\mathfrak{p}_\mu} \phi_\nu = \sum_r m_r \tilde{\omega}_{r\,\nu}. \tag{13}$$

By Lemma 8.4 we can find a differential θ with residue n_μ at $\mathfrak{p}_\mu$; denote its periods by γ_r. By Lemma 8.1 we have

$$\int \phi_\nu \wedge \theta = 2\pi i \,.\, \sum n_\mu \int_{\mathfrak{o}}^{\mathfrak{p}_\mu} \phi_\nu = \sum \gamma_r \tilde{\omega}_{r\,\nu}, \tag{14}$$

and a comparison with equation (13) shows that

$$\sum (\gamma_r - 2\pi i m_r) \tilde{\omega}_{r\,\nu} = 0.$$

Writing $x = (\gamma_1 - 2\pi i m_1, \ldots, \gamma_{2g} - 2\pi i m_{2g})$, we can express this equation as $xJ\Omega = 0$. Hence $x = y\Omega^{\mathrm{T}}$, i.e. there exists $y = (\beta_1, \ldots, \beta_g)$ such that

$$\gamma_r - 2\pi i m_r = \sum \omega_{r\,\nu} \beta_\nu.$$

Let us write $\eta = \theta - \Sigma \beta_\nu \phi_\nu$; this has the same poles and residues as θ because the ϕ_ν are of the first kind. To find its periods, we have

$$\int_{a_r} \eta = \int_{a_r} \theta - \sum_\nu \beta_\nu \int_{a_r} \phi_\nu = \gamma_r - \sum \omega_{r\,\nu} \beta_\nu = 2\pi i m_r.$$

Thus η has $2\pi i$ times integers as periods and if we set $z(\mathfrak{p}) = \exp \int_{\mathfrak{o}}^{\mathfrak{p}} \eta$, then $z(\mathfrak{p})$ is single-valued, i.e. a function on the Riemann surface and $dz/z = d(\log z) = \eta$ is meromorphic with the right poles and zeros. This shows that $\ker S = F$.

To complete the proof we must show that S is surjective. Since J is a connected complex manifold, in fact a Lie group, we need only show that a neighbourhood of 0 is covered. For let $x \in J$ and suppose that $V = \mathrm{im} S$ is a neighbourhood of 0. Then $(1/n)x \in V$ for large enough n, say $(1/n)x = S(\mathfrak{a})$ and so $S(\mathfrak{a}^n) = nS(\mathfrak{a}) = x$.

We shall need a lemma on non-special divisors:

Lemma 8.6

In a function field of genus $g > 0$ there are g distinct places $\mathfrak{p}_1, \ldots, \mathfrak{p}_g$ such that the product $\mathfrak{p}_1 \ldots \mathfrak{p}_g$ is non-special.

Proof

We have $\delta(\mathfrak{e}) = g$. Now take $\omega_1 \in D(\mathfrak{e})$, $\omega_1 \neq 0$, and choose $\mathfrak{p}_1$ to be not a zero of ω_1. Then $D(\mathfrak{p}_1) \subset D(\mathfrak{e})$, so $\delta(\mathfrak{p}_1) \leqslant g-1$. Next pick $\omega_2 \in D(\mathfrak{p}_1)$, $\omega_2 \neq 0$ and take a place $\mathfrak{p}_2$ different from $\mathfrak{p}_1$ and the zeros of ω_2; then $\delta(\mathfrak{p}_1\mathfrak{p}_2) \leqslant g-2$. Continuing in this way we reach $\delta = 0$ in at most g steps. If $\mathfrak{p}_1 \ldots \mathfrak{p}_h$ is non-special for some $h < g$, we can choose $\mathfrak{p}_{h+1}, \ldots, \mathfrak{p}_g$ arbitrarily (but distinct) to obtain a non-special divisor $\mathfrak{p}_1 \ldots \mathfrak{p}_g$.■

Now the rest of Theorem 8.5 will follow from:

Theorem 8.7 (*Jacobi*)

In a function field of genus g, let $\mathfrak{p}_1, \ldots, \mathfrak{p}_g$ be distinct places such that $\mathfrak{a} = \mathfrak{p}_1 \ldots \mathfrak{p}_g$ is non-special. Then the map

$$(\mathfrak{q}_1, \ldots, \mathfrak{q}_g) \mapsto \left(\sum_\mu \int_{\mathfrak{p}_\mu}^{\mathfrak{q}_\mu} \phi_1, \ldots, \sum_\mu \int_{\mathfrak{p}_\mu}^{\mathfrak{q}_\mu} \phi_g\right) \tag{15}$$

is a homeomorphism between a neighbourhood of $(\mathfrak{p}_1, \ldots, \mathfrak{p}_g)$ and a neighbourhood of 0 in $\mathbf{C}^g$.

We note that the right-hand side of (15) just represents

$$S\left(\frac{\mathfrak{p}_1 \ldots \mathfrak{p}_g}{\mathfrak{q}_1 \ldots \mathfrak{q}_g}\right)$$

so Theorem 8.5 follows from Theorem 8.7 and the remark preceding it.

Proof

Let t_μ be a uniformizer at $\mathfrak{p}_\mu$ and express the differential ϕ_ν near $\mathfrak{p}_\mu$ as

$$\phi_\nu = \sum h_{\nu\mu}(t_\mu)dt_\mu \quad \text{in a neighbourhood } V_\mu \text{ of } \mathfrak{p}_\mu.$$

Write $g_{\nu\mu}(t_\mu) = \int_0^{t_\mu} h_{\nu\mu}(\tau)d\tau$; then the map (15) is represented by

$$(\mathfrak{q}_1, \ldots, \mathfrak{q}_g) \mapsto \left(\sum g_{1\mu}(t_\mu), \ldots, \sum g_{g\mu}(t_\mu)\right)$$

$$= (H_1(t_1, \ldots, t_g), \ldots, H_g(t_1, \ldots, t_g)), \text{ say.}$$

Here each H_ν is holomorphic in $t_1, \ldots, t_g$ and to check that (15) is a homeomorphism near 0 we need only show that the Jacobian is non-zero. Now $\partial H_\nu/\partial t_\mu = h_{\nu\mu}(t_\mu)$, so the Jacobian at 0 is

$$\det(h_{\nu\mu}(0)) = \det\left(\frac{\phi_\nu}{dt_\mu}(\mathfrak{p}_\mu)\right). \tag{16}$$

Consider the homomorphism from $D(\mathfrak{e})$ to $\mathbf{C}^g$ given by

$$\phi \mapsto \left(\frac{\phi}{dt_1}(\mathfrak{p}_1), \ldots, \frac{\phi}{dt_g}(\mathfrak{p}_g)\right). \tag{17}$$

Both these spaces have dimension g. If ϕ maps to 0, then $\phi \in D(\mathfrak{a})$, but $\mathfrak{a}$ is non-special, so $\phi = 0$ and (17) is injective, therefore it is an isomorphism. Its determinant relative to the basis $\phi_1, \ldots, \phi_g$ is just the right-hand side of (16); so this is not zero, and this shows (15) to be a homeomorphism.■

5

Algebraic function fields in two variables

Just as an algebraic function field of one variable represents a curve, so one of two variables represents a surface. While a study of such fields would go well beyond the limits of this book, it is relatively simple to classify the valuations arising in this case, and this also illuminates the situation in the one-dimensional case. This classification of valuations is based on a paper by Zariski [1939], in which he uses these results to give a reduction of singularities of algebraic surfaces.

5.1 VALUATIONS ON FUNCTION FIELDS OF TWO VARIABLES

Throughout this section k will be an algebraically closed field of characteristic 0.

Consider a field of algebraic functions of two variables, i.e. a field E finitely generated over k and of transcendence degree 2. Thus we can find elements x, y in E algebraically independent over k, such that on writing $E_0 = k(x, y)$, we have $[E : E_0] = n < \infty$. Our aim will be to classify the valuations on E/k. Since k is algebraically closed, it must be the precise field of constants. In contrast to the one-variable case where all valuations were principal, representing points on the curve ('places'), we shall now find different kinds of valuations.

We begin with some general remarks. Let v be a valuation on E/k, with residue class field $\overline{E}$ and value group Γ. By Theorem 2.2.7 we have

$$\text{tr . deg . } \overline{E}/k + rk\,\Gamma \leqslant \text{tr . deg . } E/k = 2. \tag{1}$$

We shall exclude the trivial case where v is trivial on the whole of E; then $\text{rk}\,\Gamma = 1$ or 2 and so the dimension of the residue class field, $\overline{r} = \text{tr . deg . } \overline{E}/k$, satisfies

$$\bar{r} \leqslant 2 - \operatorname{rk} \Gamma. \tag{2}$$

We shall say that v is *$\bar{r}$-dimensional*; so we shall need to deal with the cases $\bar{r} = 0, 1$. Geometrically speaking, an $\bar{r}$-dimensional valuation on E defines an $\bar{r}$-dimensional subvariety of our surface; for $\bar{r} \leqslant r - 2$ there may be points in the neighbourhood of this subvariety, so when $r = 2$, this can only arise for $\bar{r} = 0$. Throughout we shall write B for the valuation ring of v in E, $\mathfrak{P}$ for its maximal ideal and $a \mapsto \bar{a}$ for the residue class map.

(i) One-dimensional valuations on E.

If tr. deg. $\bar{E}/k = 1$, then by (1), Γ necessarily has rank 1. We show that v must then be principal.

Proposition 1.1
Let E/k be an algebraic function field in two variables. Then any one-dimensional valuation on E/k is principal.

Proof
Let us write B_0 for the valuation ring and $\mathfrak{P}_0$ for the maximal ideal in E_0. We may assume that the transcendence basis $\{x, y\}$ lies in B_0 (replacing x by x^{-1} and similarly for y, if necessary). Then

$$R = k[x, y] \subseteq B_0.$$

Moreover, $\bar{x}, \bar{y}$ do not both lie in k, by dimensionality, say $\bar{x} \notin k$. Then x is not congruent to a constant mod $\mathfrak{P}_0$ and the prime ideal $\mathfrak{p} = \mathfrak{P}_o \cap R$ is not maximal (by the Hilbert Nullstellensatz, cf. A.2, Theorem 9.10.3). By a dimension argument (A.2, Corollary 9.10.2) it follows that $\mathfrak{p}$ is a minimal prime ideal. Now $\bar{x}, \bar{y}$ are algebraically dependent over k; if f is an irreducible polynomial such that $f(\bar{x}, \bar{y}) = 0$, then $f(x, y) \in \mathfrak{p}$ and since R is a unique factorization domain (UFD), (f) is a prime ideal (in a UFD every irreducible element is prime, cf. A.1, 6.5 and A.2, 9.2), so $\mathfrak{p} = (f)$. Hence $\mathfrak{P}_o = B_o f$, i.e. any fraction g/h in B_0 lies in $\mathfrak{P}_0$ precisely when f divides g but not h. It is clear that on B_0, $v(g/h) = m$ if and only if $g = g_o f^m$, where $f \nmid g_o h$. Hence v is principal on E_0 and so also on the finite extension E, by Theorem 2.1.1.■

We can interpret this situation geometrically as an algebraic curve on the surface corresponding to the field E, viz. the curve defined by $f(\bar{x}, \bar{y}) = 0$.

(ii) Zero-dimensional valuations on E.

In this case the rank of Γ can be 1 or 2, and when rk $\Gamma = 1$, the valuation may be non-discrete. We shall take these cases in turn.

(ii.a) We begin with the case rk $\Gamma = 2$.

Proposition 1.2

Let E/k be an algebraic function field in two variables (k algebraically closed) and let v be a valuation of rank 2 on E/k. Then v is discrete, with residue class field k, and v is composed of a principal valuation on E, followed by a principal valuation $\bar{v}_1$ on the residue class field $\bar{E}_1$ which is an algebraic function field of one variable.

Proof

By (1), v must be zero-dimensional, so Γ has just one non-trivial convex subgroup Γ_1 and the natural homomorphism $\Gamma \to \Delta = \Gamma/\Gamma_1$, when combined with $v : E^{\times} \to \Gamma$, defines a rank 1 valuation v_1 on E. Let B_1 be the corresponding valuation ring, $\mathfrak{P}_1$ the maximal ideal and $\bar{E}_1$ the residue class field. By definition we have

$$B_1 = \{a \in E \mid v(a) \geqslant 0 \text{ or } v(a) \in \Gamma_1\};$$

hence $B_1 \supset B$, $\mathfrak{P}_1 \subset \mathfrak{P}$. Any element of $B_1 \backslash \mathfrak{P}_1$ has a value in Γ_1, so we have a rank 1 valuation $\bar{E}_1 \to \Gamma_1$. Now tr.deg. $\bar{E}_1/k \leqslant 1$, and here equality holds because $\bar{E}_1$ itself has a non-trivial valuation over k; thus v_1 is a one-dimensional valuation of E. By Proposition 1.1 the value group of v_1 must be discrete and so v_1 is principal. Therefore Γ is an extension of $\mathbf{Z}$ by $\mathbf{Z}$, hence discrete of rank 2.■

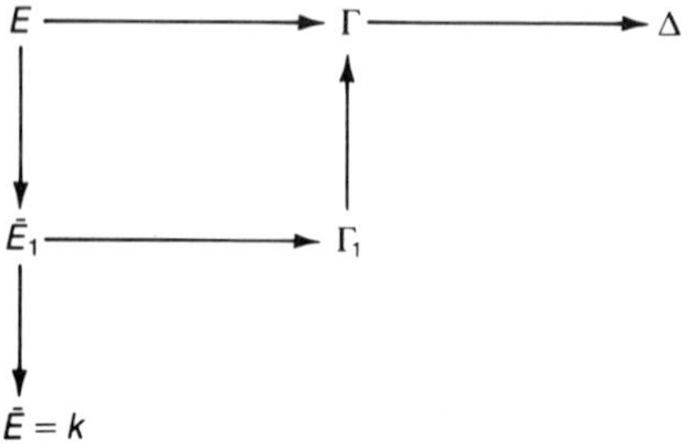

Geometrically this valuation defines a place on an algebraic curve on the surface corresponding to E. To obtain an explicit form of v, let B be the valuation ring of v on E, with prime ideals $\mathfrak{P} \supset \mathfrak{P}_1 \supset 0$ and in B choose a transcendence basis x, y for E over k and put $A = k[x, y]$, $\mathfrak{p} = \mathfrak{P} \cap A$, $\mathfrak{p}_1 = \mathfrak{P}_1 \cap A$ (more generally, A could be taken to be any integrally closed domain having E as field of fractions). If we choose $\omega \in \mathfrak{p} \backslash \mathfrak{p}_1$, then any $\xi \in E$ has the form

$$\xi = \omega^m \cdot \frac{\alpha}{\beta}, \quad \text{where } \alpha, \beta \in A \backslash \mathfrak{p}.$$

If $\bar{\alpha}, \bar{\beta}$ are the corresponding elements of $\bar{E}_1$ and $\bar{v}_1(\bar{\alpha}/\bar{\beta}) = n$, we put

$v(\xi) = (m, n) \in \mathbf{Z}^2$. Here $\mathbf{Z}^2$ with the lexicographic ordering is the value group Γ. We see that n depends on the choice of ω but m is independent of this choice. If we replace ω by ω' and the value of ω' in the above construction is $(1, \nu)$, then in terms of ω' we have

$$v(\xi) = (m', n'), \quad \text{where } m = m', n = \nu m' + n'.$$

The change of variable from ω to ω' corresponds to the automorphism

$$\begin{pmatrix} 1 & \nu \\ 0 & 1 \end{pmatrix}$$

of Γ.

There now remains the case of a zero-dimensional rank 1 valuation; this may or may not be discrete.

(ii.b) Discrete zero-dimensional rank 1 valuation.

Since the value group Γ is discrete of rank 1, we may take it to be $\mathbf{Z}$. Further, the residue class field is $\overline{E} = k$, because k is algebraically closed. Let us choose $\xi \in E$ such that $v(\xi) = 1$. Given any $\eta \in E$, if $v(\eta) = \nu$, then $v(\eta/\xi^\nu) = 0$, so there exists $c \in K$ such that $v(\eta/\xi^\nu - c) > 0$, i.e.

$$\eta = c\xi^\nu + \eta_1, \quad \text{where } v(\eta_1) = \nu_1 > \nu.$$

Repeating this process, we obtain a power series expansion

$$\eta = c\xi^\nu + c_1\xi^{\nu_1} + \ldots, \quad \nu < \nu_1 < \ldots. \tag{3}$$

On the right we have a formal power series, or if $\nu < 0$, a formal Laurent series. Writing $k((\xi))$ for the field of all such series, we have a mapping $E \to k((\xi))$, by the uniqueness of the expression (3). This is a homomorphism and hence an embedding. We thus find:

Proposition 1.3
Let E/k be an algebraic function field in two variables (k algebraically closed). Then any zero-dimensional discrete rank 1 (i.e. principal) valuation on E/k defines an embedding of E in the field of formal Laurent series $k((\xi))$. ■

Geometrically such a valuation represents a non-algebraic curve on the surface (the analytic case). For example, take the rational function field $k(x, y)$ and for any $f \in E$ define $v(f)$ as the order in t of $f(t, e^t)$. This defines a zero-dimensional principal valuation on E, corresponding to the curve $y = e^x$. Here we use the fact that t and e^t are algebraically independent over k.

(ii.c) Non-discrete zero-dimensional rank 1 valuation.

Here the value group Γ is a non-cyclic subgroup of $\mathbf{R}^+$; we distinguish two cases, depending on whether Γ is rational or not.

(ii.c′) Γ irrational
In the case of a non-rational value group we choose $x, y \in E$ such that $v(x), v(y)$ are positive incommensurable, so after normalization we may take $v(x) = 1, v(y) = \tau$, where τ is irrational. We have

$$v(x^i y^j) = i + j\tau,$$

and this shows that distinct monomials have distinct values. Given $f = \Sigma\, a_{ij} x^i y^j$, we have $v(f) = \alpha + \beta\tau$, where $x^\alpha y^\beta$ is the term of least value occurring in f. This shows firstly that x and y are algebraically independent over k, so that E is algebraic over $k(x, y)$; secondly we see that the value group Γ_0 of v on $k(x, y)$ is free abelian, generated by $1, \tau$. Now $(\Gamma : \Gamma_0) = n < \infty$ and $n\Gamma \subseteq \Gamma_0$, hence Γ, which is isomorphic to $n\Gamma$, is again free abelian of rank 2. If Γ is generated by

$$\delta_i = r_i + s_i\tau \quad (r_i, s_i \in \mathbf{Q}, i = 1, 2),$$

then by normalization we may take as generating set $1, \tau^* = \delta_2/\delta_1$. Now Γ has the basis $1, \tau^*$.

Geometrically we have a branch of the curve $y = x^\tau$. Summing up, we have:

Proposition 1.4
Let E/k be an algebraic function field of two variables (k algebraically closed). Then any non-rational rank 1 valuation on E/k has as value group a free abelian group of rank 2. ■

(ii.c″) Γ rational
Finally we come to zero-dimensional rational valuations on E/k.

Here the value group Γ consists of all rational numbers whose denominators divide some supernatural number N (Proposition 1.6.7). Since Γ is not discrete, N cannot be finite. To construct the valuation we take x, y algebraically independent in the maximal ideal $\mathfrak{P}$ of the valuation, say $v(x) = 1, v(y) = m_1/n_1$. Then $v(x^{m_1}/y^{n_1}) = 0$, so $x^{m_1}/y^{n_1} \in k$ and we have $y = c_1 x^{m_1/n_1} + y_1$, where $v(y_1) > v(y)$. By repeating the process we obtain a fractional power series

$$y = c_1 x^{m_1/n_1} + c_2 x^{m_2/n_2} + \ldots, \quad m_1/n_1 < m_2/n_2 < \ldots. \tag{4}$$

It follows from Puiseux's theorem (Corollary 4.1.2) that every algebraic function can be represented in this way; of course x and y related as in (4) need not be algebraically dependent. In fact any fractional power series (4) defines a valuation of this form and we see that by suitable choice of the denominators in (4) any supernatural number can occur in the definition of the value group.

Exercises

1. Examine how far the results of this section are valid in prime characteristic.
2. Show that in case (ii.c''), Γ is non-discrete precisely when N is not a natural number and that x, y related as in (4) are algebraically dependent only when N is a natural number.

Bibliography

The list that follows includes besides works referred to in the text books and papers which served as sources and in which the subject is developed further. The references in the text are by author's name and date (round brackets for books, square brackets for papers). For algebraic results used without proof the reader is generally referred to the author's algebra text, the volumes being cited as A.1,2 or 3, followed (or preceded) by the section number.

BOOKS

Ahlfors, L.V. (1966) *Complex Analysis*, 2nd edn, McGraw-Hill.

Artin, E. (1956) *Theory of Algebraic Numbers*, Notes by G. Würges, Göttingen.

Behnke, H. and Sommer, F. (1962) *Theorie der analytischen Funktionen einer komplexen Veränderlichen* (2. *Aufl. Grundlehren d. math. Wiss.* 77), Springer-Verlag.

Borevich, Z.I. and Shafarevich, I.R. (1966) *Theory of Numbers* (in Russian) 2nd edn. Izdat. Nauka Moskva, 1972 (Translation of 1st edn in *Pure and Applied Mathematics*, Vol 20, Academic Press).

Bourbaki, N. (1966) *General Topology*, ch. 2, Addison-Wesley.

Bourbaki, N. (1972) *Commutative Algebra*, ch. 6, Addison-Wesley.

Chevalley, C. (1951) *Introduction to the Theory of Algebraic Functions of One Variable* (*Math. Surveys*, No. VI), Amer. Math. Soc., New York.

Cohn, P.M. (1982, 1989, 1990) *Algebra*, 2nd edn. Vols 1–3, J. Wiley, Chichester. Cited in the text as A.1, A.2, A.3 respectively.

Eichler, M. (1966) *Introduction to the Theory of Algebraic Numbers and Functions* (*Pure and Applied Mathematics*, Vol. 23), Academic Press.

Hartshorne, R. (1977) *Algebraic Geometry* (*Graduate Texts in Mathematics*, Vol. 52), Springer-Verlag.

Hasse, H. (1980) *Number Theory* (*Grundl. d. math. Wiss.*, Vol. 229), Springer-Verlag.

Hensel, K. and Landsberg, G. (1902) *Theorie der algebraischen Funktionen einer Variablen*, G. B. Teubner. Chelsea Reprint 1965.

Jung, H. W. E. (1923) *Algebraische Funktionen einer Veränderlichen*, De Gruyter.

Jung, H. W. E. (1951) *Einführung in die Theorie der algebraischen Funktionen zweier Veränderlicher*, Akademie Verlag, Berlin.

Lang, S. (1970) *Algebraic Number Theory*, Addison-Wesley.

Lefschetz, S. (1949) *Introduction to Topology* (*Princeton Math. Series* 11), Princeton University Press.

Matzat, B. H. (1987) *Konstruktive Galois-Theorie* (*Lecture Notes in Mathematics*, 1284), Springer-Verlag.

Picard, E. (1893) *Traité d'Analyse* (3 vols), Gauthier-Villars, Paris.

Rudin, W. (1966) *Real and Complex Analysis*, McGraw-Hill.

Schilling, O. F. G. (1950) *The Theory of Valuations (Math. Surveys* IV), Amer. Math. Soc., New York.

Semple, J. G. and Roth, L. (1949) *Introduction to Algebraic Geometry*, Clarendon Press, Oxford, 1987.

Schafarewitsch, I. R. (ed.) (1968) *Algebraische Flächen*, Akademische Verlagsgesellschaft, Leipzig.

Tchebotarev, N. G. (1948) *Theory of Algebraic Functions* (in Russian), OGIZ, Moscow.

van der Waerden, B. L. (1939) *Einführung in die algebraische Geometrie*, Dover Reprint 1945.

Weil, A. (1967) *Basic Number Theory (Grundl. d. math. Wiss.*, Vol. 144), Springer-Verlag.

PAPERS

Artin, E. [1932] Über die Bewertungen algebraischer Zahlkörper. J. reine u. angew. Math., 167, 157–9. (Collected Papers, pp. 199–201, Addison-Wesley, 1965.)

Artin, E. and Whaples, G. [1945] Axiomatic characterization of fields by the product formula for valuations. *Bull. Amer. Math. Soc.*, 51, 469–92. (*Collected Papers*, pp. 202–25.)

Chevalley, C. [1940] La théorie des corps de classes. *Ann. of Math.*, **41**, 394–418.

Claborn, L. [1966] Every abelian group is a class group. *Pacif. J. Math.*, **18**, 219–22.

Clifford, W. K. [1878] On the classification of loci. *Phil. Trans. Roy. Soc.*, **169**, 663–81. (*Collected Papers*, pp. 305–31. Chelsea Reprint 1968.)

Dedekind, R. and Weber, H. [1879] Theorie der algebraischen Funktionen einer Veränderlichen. *J. reine u. angew. Math.*, **92**, 181–290.

Golod, E. S. and Shafarevich, I. R. [1964] On class field towers (in Russian).

Izvestiya Akad. Nauk SSSR, Ser. Mat. **28**, 261–72. (*Amer. Math. Soc. Transl.* Ser. 2, **48** (1965), 91–102.)

Harnack, A. [1876] Ueber die Vieltheiligkeit der ebenen algebraischen Curven. *Math. Ann.*, **10**, 189–98.

Hurwitz, A. [1893] Über algebraische Gebilde mit eindeutigen Transformationen in sich. *Math. Ann.*, **41**, 403–42. (*Math. Werke I*, pp. 391–430, Basel, 1932.)

Krull, W. [1931] Allgemeine Bewertungstheorie. *J. reine u. angew. Math.*, **167**, 160–96.

Ostrowski, A. [1917] Über einige Lösungen der Funktionalgleichung $\phi(x)\phi(y) = \phi(xy)$. *Acta Math.*, **41**, 271–84.

Ostrowski, A. [1934] Untersuchungen zur arithmetischen Theorie der Körper. *Math. Zeits.*, **39**, 269–404.

Rychlik, K. [1923] Zur Bewertungstheorie der algebraischen Körper. *J. reine u. angew. Math.*, **153**, 94–108.

Zariski, O. [1939] The reduction of singularities of an algebraic surface. *Ann. of Math.*, **40**, 639–89 (*Collected Papers*, I, pp. 325–75, MIT Press, 1972.)

Table of notations

Symbol	Meaning	Page
$\lvert x \rvert$	Absolute value of x	1
$\hat{R}$	Completion of R	12
$\lVert x \rVert$	Norm of x (in a normed space)	14
$\mathbf{Q}_p$	Field of p-adic numbers	26
$\mathbf{Z}_p$	Ring of p-adic integers	29
$k((t))$	Field of formal Laurent series	30
$\mathrm{rk}\,\Gamma$	Rank of ordered group Γ	32
f	Residue degree	43
e	Ramification index	43
$N_{L/K}(\alpha)$	Norm of α	50
$\mathrm{tr}\,.\,\mathrm{deg}\,(L/K)$	Transcendence degree of L over K	51
$E \otimes F$	Tensor product of E and F	54
$E \times F$	Direct product of E and F	55
$Tr_{L/K}(\alpha)$	Trace of α	56
$D(U)$	Discriminant of A-module U	76
$\mu(\mathfrak{A})$	Volume of parallelotope $\mathfrak{A}$	94
$L(\mathfrak{A})$	Subset of elements bounded by a parallelotope $\mathfrak{A}$	94
	Subspace of functions with divisor a multiple of $\mathfrak{A}$	115
$\lambda(\mathfrak{A})$	Order of parallelotope $\mathfrak{A}$	95
$d(\mathfrak{A})$	Degree of a divisor $\mathfrak{A}$	115
$\dim_k(V)$, $[V:k]$	Dimension of k-space V	116
(u)	Divisor of a function u (also, the ideal generated by u)	118
$(u)_0$	Divisor of zeros of u	118
$(u)_\infty$	Divisor of poles of u	118
D	Divisor group	114
D_0	Group of divisors of degree 0	115
$l(\mathfrak{A})$	Dimension of $L(\mathfrak{A})$	116

Symbol	Meaning	Page
J	Jacobian variety	123, 171
C	Divisor class group	69
$N(C)$	Dimension of divisor class	123
g	Genus	125
$\delta(\mathfrak{a}^{-1})$	Specialty index	126
Λ	Adele ring	126
$res_{\mathfrak{p}}(\omega)$	Residue of ω at $\mathfrak{p}$	140
$D(\mathfrak{a})$	Space of differentials divisible by $\mathfrak{a}$	142
W	Canonical divisor class	139
E	Principal divisor class	145
Ω	Period lattice	155, 161
$\wedge$	Wedge product of differentials	168
P	Picard variety	171

Index